Managing Technology Transition in Saudi Arabia

Khalid A. Alrashoud

Managing Technology Transition in Saudi Arabia

Residential Solar Photovoltaic Systems Development

Springer

Khalid A. Alrashoud
Department of Transdisciplinary
Science and Engineering
Tokyo Institute of Technology
Tokyo, Japan

ISBN 978-981-15-5802-3 ISBN 978-981-15-5803-0 (eBook)
https://doi.org/10.1007/978-981-15-5803-0

This Springer imprint is published by the registered company Springer Nature Singapore Pte Ltd.
The registered company address is: 152 Beach Road, #21-01/04 Gateway East, Singapore 189721, Singapore

To the soul of the late King Abdullah bin Abdulaziz Al Saud, the Custodian of the Two Holy Mosques, who ordered to re-launch the ambitions Overseas Scholarship Program in 2005, allowing hundreds of thousands of enthusiastic Saudi youth, including myself, to pursue their dreams in acquiring higher academic degrees and assist in building their country.

To King Salman Bin Abdulaziz Al Saud, the Custodian of the Two Holy Mosques, and the Prime Minister, who continues to support science and knowledge by empowering his people.

To His Royal Highness Prince Mohammad bin Salman bin Abdulaziz Al Saud, the Crown Prince, First Deputy Prime Minister, Minister of Defense, and Chairman of the Council of Economic and Development Affairs. The reformer, who inspires Saudis to place their country where it deserves.

Preface

Amidst the depletion of fossil fuel reserves, unsustainable domestic energy consumption, Greenhouse Gas (GHG) emissions, along with the increasing and rapid population growth have given stimulation for Saudi Arabia to shift to more sustainable energy systems. In order to assist a sustainability transition in the energy sector, involved actors need robust knowledge about the obstacles and drivers to the implementation of clean energy technologies. Solar Photovoltaics (PV) system is one of the most technically mature and economically competitive technology in many parts of the world. Not least at the residential level, PV has significant potential. Although Residential Solar Photovoltaic Systems (RSPSs) have reached economic competitiveness, this technology might still confront significant obstacles in the socio-technical systems in which it is being deployed.

In addition to the recent development of the utility-scale solar energy projects, in 2018, the Saudi government called upon individuals wishing to integrate RSPSs at their residence units to submit their application. However, the number of applications received is significantly small and falling short of expectations. To address this issue, a comprehensive multi-faceted exploration in its real-life settings is required.

This book aims at advancing knowledge regarding the current technology transition of RSPSs in Saudi Arabia. It is largely based on various forms of data collection and analysis which were performed in Saudi settings at the societal and state levels. It applies the *Diffusion of Innovations* theory (*DOI*) targeting households to examine factors that may potentially motivate or impede them from adopting RSPSs. This research also employs the *Multi-level Perspective* (*MLP*) model as an analytical tool to understand the dynamics of such transition by analyzing the factors shaping the RSPSs policies through exploring the broader energy systems.

Tokyo, Japan

Khalid A. Alrashoud

Acknowledgments

To my life-guide, my dear mother, to whom I owe everything, many Thanks! To my other family members who have supported me along the way. This book would not have been possible without their warm love, continued patience, and endless encouragement. To my late uncle Khalid who believed in me and taught me that hard work pays off and that so much could be done with little.

A very special appreciation and gratitude go to Prof. Tokimatsu Koji for his endless support. His advice has been invaluable. Under his supervision, I learned how to improve my academic skills. Through the hard times, I was always inspired by his hardworking, commitment, and passionate attitude. I would also like to thank Prof. Takahashi Fumitake for his encouragement and helpful remarks. His vibrant soul and sense of humor have always been a source of joy and inspiration.

Thanks also go to Profs. Yuya Kajikawa, Shigeo Nishikizawa, and Daisuke Akita for their insightful comments and guidance. Their crucial remarks stemming from long experience have shaped the final version of this book.

Contents

1 Introduction 1
1.1 Background Outline 1
1.1.1 Research Background 1
1.1.2 Brief Historical Perspective on Solar Energy in Saudi Arabia 2
1.1.3 Rationale for Research 3
1.2 Previous Literature 6
1.3 Previous Research Gaps 13
1.4 Research Objectives, Questions, and Hypotheses 15
1.4.1 Research Objectives 15
1.4.2 Research Questions 15
1.4.3 Research Hypotheses 16
1.5 Research Structure, Research Interconnected Relationships, and Research Originality 17
1.5.1 Research Structure 17
1.5.2 Research Interconnected Relationships 18
1.5.3 Research Originality 20
References 20

2 Global Practices: Assessing Renewable Energy Policies 23
2.1 Definition of Terms 23
2.2 Methodology and Research Design 24
2.3 Results and Discussion 25
2.3.1 Renewable Energy Policies in Saudi Arabia 25
2.3.2 Renewable Energy Policies in Germany 32
2.3.3 Renewable Energy Policies in Japan 34
2.3.4 Renewable Energy Policies in the USA 35
2.3.5 Renewable Energy Policies in the UK 37
2.3.6 Renewable Energy Policies in Spain 38

2.3.7 Renewable Energy Policies in India 40
2.3.8 Renewable Energy Policies in Greece 42
2.4 Conclusion 44
2.5 Limitations 46
References 47

3 An Exploratory Study of the Public's Views on Residential Solar Photovoltaic Systems in Hydrocarbon-Rich Saudi Arabia 49
3.1 Research Background 49
3.2 Social Acceptance of RSPSs 51
3.3 Research Questions and Hypotheses 53
3.3.1 Research Question 53
3.3.2 Research Hypotheses 53
3.4 Methodology and Research Design 53
3.4.1 Sampling Method 54
3.4.2 Questionnaire Design 55
3.5 Results 55
3.5.1 Basic Demographic Characteristics 55
3.5.2 Residence Status 59
3.5.3 Household Energy Use 59
3.5.4 Environmental and Energy Cognition 61
3.5.5 Willingness to Adopt RSPSs 65
3.5.6 Multivariate Analysis 65
3.6 Discussion 69
3.7 Recommendations 72
3.8 Limitations 73
3.9 Conclusion 73
References 74

4 Critical Factors Influencing Social Perception of Residential Solar Photovoltaic Systems in the Hydrocarbon-Rich Saudi Arabia 77
4.1 Research Background 77
4.2 Diffusion of Innovations Framework 80
4.2.1 The Mechanism of Innovation Diffusion 80
4.2.2 Consumer Behavior and Innovation Diffusion 82
4.3 Research Questions and Hypotheses 83
4.3.1 Research Questions 83
4.3.2 Research Hypotheses 83
4.4 Methodology and Research Design 84
4.4.1 Questionnaire Design 85
4.4.2 Innovation Attributes and Factors Selection 86
4.4.3 Date Analysis 89
4.5 Results 90

4.5.1 Demographic Status . . . 90
4.5.2 Solar Energy Cognition . . . 92
4.5.3 Innovation Attributes Assessment . . . 93
4.6 Discussion . . . 97
4.7 Recommendations . . . 100
4.8 Limitations . . . 102
4.9 Conclusion . . . 102
References . . . 103

5 Multi-level Perspective of Technological Transitions: Analysis of the Residential Solar Photovoltaic Systems Policies in Hydrocarbon-Rich Saudi Arabia . . . 107
5.1 Research Background . . . 107
5.2 Socio-Technical Transitions to Sustainability . . . 108
5.3 Conceptual Framework: Multi-level Perspective and Policy Design Modeling of Transition Management . . . 109
5.4 Research Question and Hypotheses . . . 118
5.4.1 Research Question . . . 119
5.4.2 Research Hypotheses . . . 119
5.5 Methodology and Research Design . . . 119
5.5.1 Interviewees Categorization . . . 122
5.6 Results and Discussion . . . 124
5.6.1 Landscape (Macro-Level) . . . 124
5.6.2 Regime (Meso-Level) . . . 136
5.6.3 Niche-Level (Micro) . . . 149
5.6.4 Summary of the Key Findings in the Analysis of Saudi Arabia Solar Energy System . . . 151
5.7 Policy Design Modeling of Transition Management . . . 154
5.8 Recommendations . . . 157
5.9 Limitations . . . 157
5.10 Conclusion . . . 158
References . . . 158

6 Discussion and Conclusion . . . 163
6.1 Results of the Study . . . 163
6.2 Evaluation of Research Hypotheses . . . 165
6.3 Overall Conclusion . . . 167
6.4 Key Contributions of the Research . . . 168
6.4.1 Framework Applicability . . . 169
6.4.2 Empirical Contribution . . . 170
6.5 Limitations and Avenues for Future Research . . . 171
6.6 Recommendations . . . 172
Annex . . . 174
References . . . 175

Correction to: An Exploratory Study of the Public's Views on Residential Solar Photovoltaic Systems in Hydrocarbon-Rich Saudi Arabia . . . C1

Abbreviations

CCPI	Climate Change Performance Index
CDM	Clean Development Mechanism
CI	Confidence Interval
CSLF	Carbon Sequestration and Leadership Forum
CSOs	Civil Society Organizations
CSP	Concentrated Solar Power
DOI	Diffusion of Innovation
DSP	Distribution Service Provider
ECRA	Electricity and Co-Generation Regulatory Authority
EEG	Binding Equalization Scheme
EVs	Electric Vehicles
FDI	Foreign Direct Investment
FiT	Feed-in-Tariff
GBI	Generation-Based Incentive
GCC	The Gulf Cooperation Council
GDP	Gross Domestic Product
GHG	Greenhouse Gas emissions
IC	Industrial Clusters
IEC	International Electrotechnical Commission
IMF	International Monetary Fund
IPP	Independent Power Producers
IRENA	The International Renewable Energy Agency
ITC	Investment Tax Credit
K.A.CARE	King Abdullah City for Atomic and Renewable Energy
KACST	King Abdulaziz City for Science and Technology
KAPSARC	King Abdullah Petroleum Studies and Research Center
KAUST	King Abdullah University of Science and Technology
KSA	The Kingdom of Saudi Arabia
LCOE	Levelized Cost of Energy
MASH	Multifamily Affordable Solar Housing Program

MCAR	Missing Completely at Random
MEIM	The Ministry of Energy, Industry and Mineral Resources
MENA	The Middle East and North Africa
MESIA	Middle East Solar Industry Association
MLP	Multi-level Perspective
MLSD	Ministry of Labor and Social Development
MNRE	Ministry of New and Renewable Energy
MOE	Ministry of Energy
NAPCC	The National Action Plan on Climate Change
NREP	The National Renewable Energy Program
PIF	Public Investment Fund
PPAs	Power Purchase Agreements
PV	Photovoltaics
R&D	Research and Development
RCJY	Royal Commission for Jubail and Yanbu
RECs	Renewable Energy Certificates
REPDO	The Renewable Energy Project Development Office
RES	Renewable Energy Sources
RET	Renewable Energy Technology
RETs	Renewable Energy Technologies
RO	Renewable Obligation
ROC	Renewable Obligation Certificate
RPO	Renewable Purchase Obligation
RPS	Renewable Portfolio Standard
SAGIA	Saudi Arabian General Investment Authority
SAREIF	The Saudi Arabia Renewable Energy Investment Forum
SASH	Single-family Affordable Solar Homes program
SASIA	Saudi Arabia Solar Industry Association
SASO	The Saudi Standards, Metrology and Quality Organization
SEC	The Saudi Electricity Company
SGBF	Saudi Green Building Forum
SMART	System-integrated Modular Advanced Reactor
SMBs	Small and Medium-sized Businesses
SNM	Standard Net Metering
SNS	Social Networking Service
ST	Socio-technical
TM	Transition Management
ToU	Time-of-Use
UAE	The United Arab Emirates
UNFCCC	The United Nations Framework Convention on Climate Change
VNM	Virtual Net Metering
WTO	World Trade Organization

List of Figures

Fig. 1.1 Correlation between Saudi Arabia GPD growth and crude oil price growth. *Source* (Sullivan 2016) 4
Fig. 1.2 Saudi Arabia carbon dioxide emissions 1960–2019. *Source* (macrotrends 2019) 5
Fig. 1.3 Photovoltaic power potential in Saudi Arabia. *Source* (SOLARGIS 2016) 6
Fig. 1.4 Research gaps and directions 15
Fig. 1.5 Schematic presentation of research framework 19
Fig. 1.6 Research interconnected relationships as a hierarchy 19
Fig. 1.7 Research specific and overall originality 20
Fig. 2.1 Saudi national renewable energy revised targets 26
Fig. 2.2 Estimated GCC countries energy subsidies as a percent of GDP 28
Fig. 2.3 Rooftop Net Metering diagram 30
Fig. 2.4 Small-scale solar PV systems connection process as proposed by ECRA 30
Fig. 3.1 Respondents' knowledge of global warming causes and effects 63
Fig. 3.2 Respondents' recommendation on global warming 63
Fig. 3.3 Respondents' decisions on paying extra money for electricity monthly bill to address the effects of global warming 63
Fig. 3.4 Respondents' expectations on the longevity of crude oil in Saudi Arabia 64
Fig. 3.5 Respondents' preferences on crude oil as a source of energy and perception of the existence of an air pollution issue in Saudi Arabia 64
Fig. 3.6 Respondents heard about solar energy for the first time 64
Fig. 3.7 Participants' willingness to install a residential solar photovoltaic system with subsidies 65
Fig. 4.1 Respondents knowledge of solar energy (n = 1498) 92

Fig. 4.2 Participant response regarding the variables of the innovation attributes. **a** Participant response on *relative advantage* variables, **b** participant response on *observability* variables, **c** participants response on *complexity* variables, **d** participants response on *compatibility* variables (In order to make the comparison between acceptors and rejectors easier, the questions in the rejectors Likert scale were recorded in reverse order to match the acceptors' scale) 95
Fig. 5.1 The multi-level perspective on transitions. *Source* Geels (2002) 110
Fig. 5.2 The five socio-technical regimes. *Source* Geels (2004). 111
Fig. 5.3 Cycles of transition management. *Source* Loorbach (2007) 117
Fig. 5.4 Renewable energy stakeholders in Saudi Arabia (2019). *Source* Own compilation. 130
Fig. 5.5 Numbers of scientific publications in Arab countries. *Source* Al-Suwaiyel (2015). 148
Fig. 6.1 Distribution of the innovation attributes scores among acceptors (left) and rejectors (right) of RSPSs 174

List of Tables

Table 1.1	A summary of previous key studies on the interaction of individuals with renewable energy	10
Table 2.1	Power generation by source in Saudi Arabia, 2016	25
Table 2.2	Incentives offered by Saudi government entities for utility-scale project developers	27
Table 2.3	Scope and application of the Small-scale solar PV systems regulations introduced by ECRA	29
Table 2.4	Electricity tariffs for all categories of electricity consumption	31
Table 2.5	List of fees for services relevant to the connection of small-scale solar PV systems	32
Table 2.6	A summary of actions taken by the government of Saudi Arabia	32
Table 2.7	A summary of actions taken by the government of Germany	34
Table 2.8	A summary of actions taken by the government of Japan	35
Table 2.9	A summary of actions taken by the government of the USA	37
Table 2.10	A summary of actions taken by the government of the UK	39
Table 2.11	A summary of actions taken by the government of Spain	41
Table 2.12	A summary of actions taken by the government of India	42
Table 2.13	A summary of effective and ineffective actions taken by the government of Greece	43
Table 2.14	Summative classification of ineffective actions taken by governments	45
Table 3.1	Frequencies and proportions of RSPS acceptance by demographic characteristics (n = 1498)	56
Table 3.2	Numbers and proportions of acceptance of RSPSs by residence status (n = 1498)	58

Table 3.3 Association between renting and the likelihood of translocating in the next ten years 59
Table 3.4 Frequencies and proportions of acceptance of RSPSs by household energy use (n = 1498) 60
Table 3.5 Correlation between monthly income and frequency of checking electricity consumption 62
Table 3.6 Binary logistic regression analysis for variables with statistical significance in the bivariate analysis (n = 907, $R^2_{Nagelkerke} = 0.463$). 66
Table 3.7 Binary logistic regression analysis (n = 907, $R^2_{Nagelkerke} = 0.259$). 68
Table 4.1 Innovation attributes and factors definition. 88
Table 4.2 Reliability testing across different attributes 90
Table 4.3 Frequencies and proportions of residential solar photovoltaic systems acceptance based on demographic characteristics (n = 1498) 91
Table 4.4 Statistics across different attributes and tests to ascertain differences in medians 94
Table 4.5 The association between attributes/variables and accepting/rejecting RSPS installation. 96
Table 4.6 Correlation between different scales used in the study 98
Table 5.1 Analytical dimensions of the RSPSs policies in hydrocarbon-rich Saudi Arabia structured 112
Table 5.2 Key modification to the analytical framework developed by Kumetat for the Hydrocarbon-rich Arab countries. 115
Table 5.3 Interviewees' backgrounds and approach methods 120
Table 5.4 Analytical dimensions structured by subsections 121
Table 5.5 Interviewees category definition 122
Table 5.6 Key findings of Saudi Arabia's solar RSPSs policy analysis 154
Table 5.7 TM policy design, realpolitik caveats, and expected outcomes 155

Chapter 1
Introduction

This chapter presents the background, objectives and, motivations of this research. The early solar energy development and rationale for the research are also presented. Further, the previous literature and the literature gaps are identified in this chapter. Lastly, the research questions, hypotheses, and the originality of this research are also stated.

1.1 Background Outline

1.1.1 Research Background

In addition to its unique position as a leading country in the Arab and Islamic worlds, The Kingdom of Saudi Arabia enjoys several exceptional and strategic geopolitical and economic advantages. Since it constitutes the majority of the Arabian Peninsula, Saudi Arabia is located at the crossroads of vital international trade routes that link three continents: Asia, Africa, and Europe. It is also the largest exporter of petroleum and has around 18% of the world's proven oil reserves. Substantial revenues derived from fossil fuel exports have been the key element in the development and prosperity of Saudi Arabia. However, due to the pressure of the depletion of oil reserves and rapid population growth, Saudi Arabia faces enormous challenges in diversifying its petroleum-based economy structures, job creation, GHG emissions reduction. As proven in many countries, the implementation of renewable energy projects can play a pivotal role in achieving these goals. Among many approaches that Saudi Arabia took recently to develop its renewable energy sector is to allow individual households to possess and integrate their residential solar photovoltaic systems (RSPSs) to the national grid, which has proceeded slowly. Applying other countries' models to the case in the Kingdom may seem difficult due to the variations in the socio-political,

K. A. Alrashoud, *Managing Technology Transition in Saudi Arabia*,
https://doi.org/10.1007/978-981-15-5803-0_1

socio-technical compositions. This book presents the situation with regard to the individuals' perception of RSPSs and the policies currently placed for its implementation. It also discusses the possible approaches to enhance the adoption of this technology.

1.1.2 Brief Historical Perspective on Solar Energy in Saudi Arabia

Since the 1960s, Saudi Arabia has considered the critical situation of depending on a finite source such as oil as the main national income. In response to the International Monetary Fund (IMF) suggestions, Saudi Arabia developed the so-called a (Five-year-Plan) to create a diversified economy and increase other sectors' contribution to the gross domestic product (GDP). The first application of solar energy can be traced back to the 1960s when a French company had established the first solar PV beacon at a small airport of Al-Madīnah Al-Munawarah city (Sakr 1987). This was followed by small-scale university experimental projects during 1969. During that time, King Abdulaziz City for Science and Technology (KACST) conducted the first systematized solar energy research and development (R&D) activities in 1977.

Since then, the Energy Research Institute (ERI) at KACST has been working on joint programs with international research and development centers, conducting R&D works in the field of solar energy. Between 1975 and 1997, under an agreement entered into between Saudi Arabia and the United States, an R&D program named 'SOLERAS' was launched with the aim to design and install several solar energy pilot projects. In the 1980s, SOLERAS installed off-grid solar applications supplying two traditional villages, Al-Uyaynah and Al-Jubaylah, which are located around 30 km northwest of the capital, Riyadh. Back then, these two projects were considered as innovation of the region, making Saudi Arabia the first country in the Gulf Cooperation Council (GCC) countries and the Middle East to install solar system independent from the public electric grid (Hepbasli and Alsuhaibani 2011).

Another important starting point of solar energy development in Saudi Arabia is the long-term cooperative program between The Federal Republic of Germany and Saudi Arabia 'HYSOLAR' in 1986. Through two phases, HYSOLAR was carried out, focusing on developing and demonstrating solar hydrogen production. Whereas the first phase mainly focused on the improvement of hydrogen production technologies, the second phase emphasis was more laid on hydrogen utilization technologies (Hepbasli and Alsuhaibani 2011). Since then, research activities have progressed slowly with no implantation of large-scale real-life solar energy projects. In 2010, the Saudi government established King Abdullah City for Atomic and Renewable Energy (K.A.CARE) to serve as a center responsible for developing the country's capabilities of renewable and atomic technologies. In 2012, K.A.CARE announced its plan to install 41 gigawatts (GW) of renewable energy by 2032; however, K.A.CARE has failed to kick off these projects.

1.1.3 *Rationale for Research*

The rationale for this research can be broadly categorized into economic, environmental, and potential for solar energy. Each of these rationales is discussed in the sequence below.

1.1.3.1 Economic Factors

Energy is a vital element for the economic growth of any society. Among various energy forms, electricity is considered the most significant source in recent times, as it underpins every other source of energy. It has been argued that, without maintaining at least the basic electricity access, no country can develop beyond a subsistence living (The World Bank 2000). This is because electricity is essential to clean water supply, cultivating food, education, job creation, healthcare, housing, and national security. Saudi Arabia is a country with rich energy sources. It is considered the world's largest oil producer, which also controls the world's second-largest oil reserves (EIA 2016).

Nevertheless, the Saudi economy has witnessed a substantial growth that is predicted to reach instability if no measures are taken. Oil revenues account for around 90% of the Saudi government's fiscal revenues, and around 85% of export revenues. Moreover, the oil sector contributes more than 40% to the GDP, which means the socio-economic status is severally subject to oil price fluctuation (Fig. 1.1).

The IMF perceives a high threat in economies that heavily depend on oil, describing fossil fuel prices as "a key source of macroeconomic volatility" (IMF 2016), putting the Saudi economy at risk to volatility in the global oil market (Jalilvand and Westphal 2017). Another issue caused by the heavy reliance on oil is lock-in opportunities to achieve further growth (Chevalier and Geoffron 2016). Limited growth can pose even more severe threats in Saudi Arabia which is facing increasing demand for energy, a growing population, and high unemployment rates.

In fact, the unsustainable increase for energy demand, in tandem with the rapid population growth, has been eating away the oil reserves rather than exporting it (Jalilvand and Westphal 2017), resulting in increasing pressure on the country's budget. Some experts argue that if the local energy consumption continues at the current levels, it could outpace oil production. This means that Saudi Arabia may become a net energy importer (Nachet and Aoun 2015), which could cause not only instability to Saudi Arabia's economy but also to the global economy.

1.1.3.2 Environmental Factors

After all, the extensive wealth which the country relies on to continue developing its infrastructure providing welfare to its citizens is based on a finite resource, which is, by nature, a passing phenomenon. Being a state with a "rentier economy" is the main challenge that defines the possibilities for the future of Saudi Arabia. Giacomo

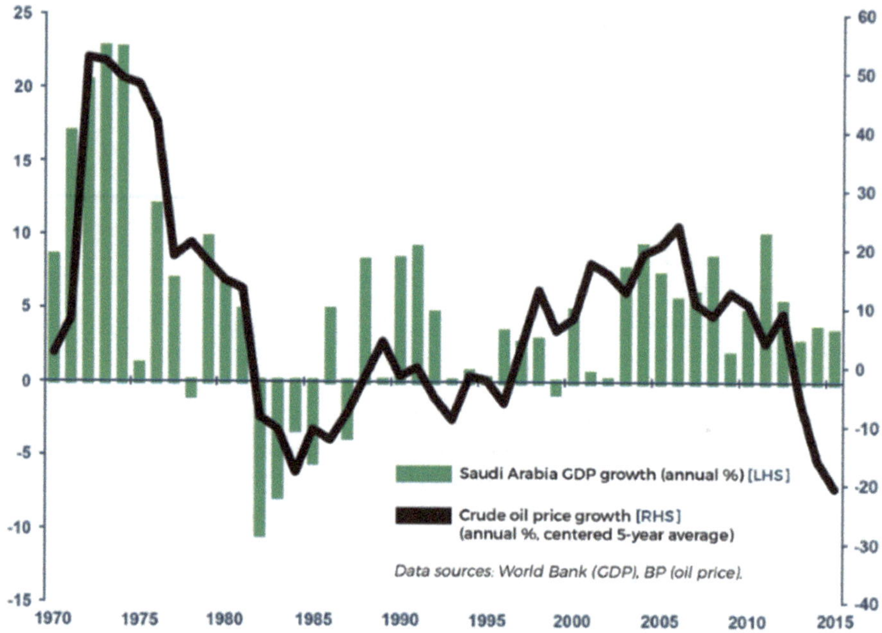

Fig. 1.1 Correlation between Saudi Arabia GPD growth and crude oil price growth. *Source* (Sullivan 2016)

Luciani suggests two scenarios for the future of rentier states: the first scenario is that a rentier state makes a fundamental change by pursuing diversification of its economic base and gradually become a production state, while the second scenario is to continue consuming the fossil fuel reserves without change until they are completely depleted and then simply disappear. The second scenario seems to be approaching faster than Luciani had predicted, setting Saudi Arabia at crossroads. This not only because of the expected depletion of the Saudi oil reserves but also due to a global reorientation shifting away from the environment harming fossil fuel sources which could weaken demand for fossil fuel (Tagliapietra 2017). While studies projected that the oil reserves in Saudi Arabia could last for more 63 years before they are physically depleted, the international agreements (e.g., Paris Agreement) for decarbonization and technological advancement could make the demand for oil uncertain by then (Tagliapietra 2017).

Moreover, Saudi Arabia is not less vulnerable when it comes to the effects of global warming. In fact, its hot-arid climate makes it highly at risk of the adverse impacts of global warming. In addition to its vulnerability to the effect of global warming, Saudi Arabia contributes substantially to carbon dioxide (CO_2) emission (601,046.97 kilotons in 2014, with an increase of 11.09% from 2013), which poses even more threats on Saudi Arabia (Fig. 1.2).

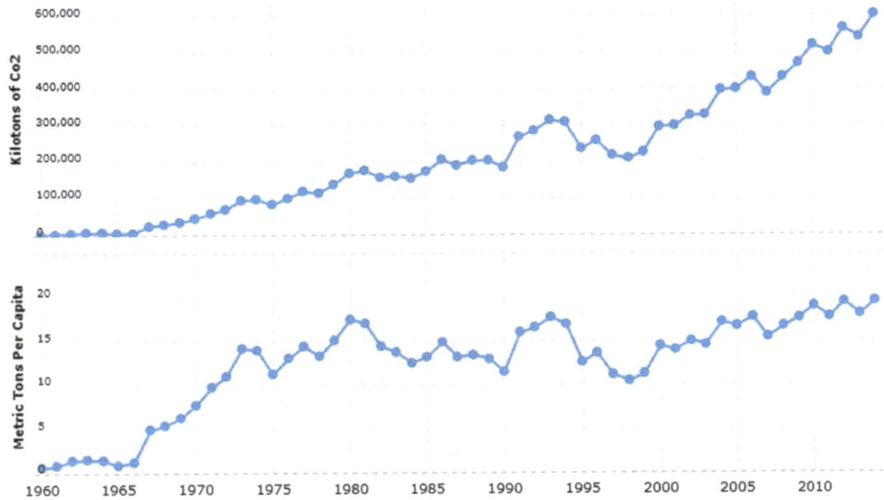

Fig. 1.2 Saudi Arabia carbon dioxide emissions 1960–2019. *Source* (macrotrends 2019)

With the increasing CO_2 emission combined with the underperformance of energy efficiency in Saudi Arabia (ranked 25th in 2018), global warming becomes a double-edged weapon for the country (ACEEE 2018). As the impact of climate change becomes more evident, the global temperature increases continuously. This will lead to increased demand for energy, which will eventually exacerbate the issue, making it even more urgent for the Kingdom to seek a diversified energy base (Luciani 2013).

1.1.3.3 Solar Energy Potential

Although Saudi Arabia may not be able to change the fact of its dependency on oil exports, it can still become an exporter of a non-oil derived energy (Mouritsen 2018). Saudi Arabia is considered to have one of the world's highest solar irradiation, estimated at 2,200 thermal kilowatt hours per square meter (kWh/m^2) (Alawaji 2001) (Fig. 1.3).

The vast territory and abundant deposits of silica sand that can be used in the manufacture of PV panels make solar energy an ideal option to diversify the energy portfolio of Saudi Arabia (Nash et al. 2017). An economic reforming plan recently launched, named "Saudi Vision 2030" is a key step in the integration of solar energy as a part of the country's energy mix. Through utility-scale solar energy stations or at distributed on-site micro-scale solar PV systems, with individuals integrating RSPSs to their homes to become co-producers, Saudi Arabia is aiming to become less dependent on fossil fuel. Although the implementation of utility-scale solar projects might seem to be less complicated, bringing tangible results in less time, the government also seeks to spread the RSPSs through the residential sector which

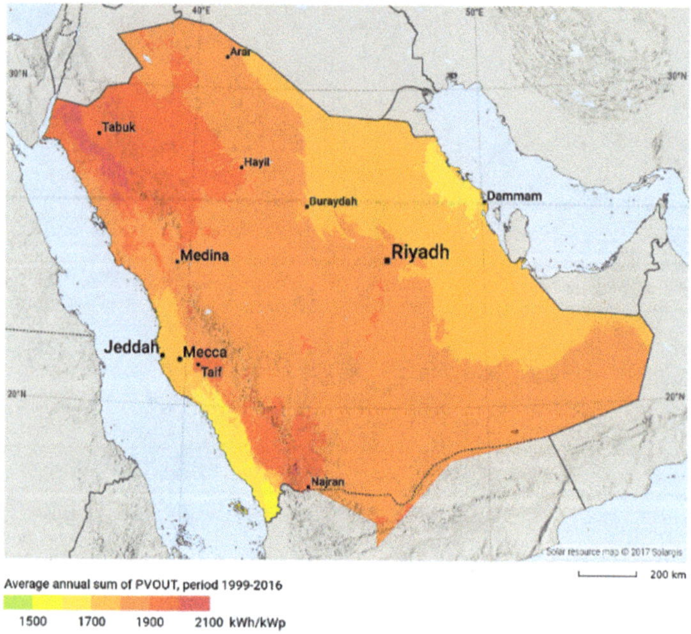

Fig. 1.3 Photovoltaic power potential in Saudi Arabia. *Source* (SOLARGIS 2016)

accounts for half of the domestic electricity consumption (Tlili 2015). A wide-range adoption of RSPSs can play a significant role in easing the pressure on the national grid as the peak load demand occurs at the same time with the high solar irradiation (Almasoud and Gandayh 2015). Nevertheless, since individuals were allowed to integrate their RSPSs to the pubic electric grid, the rate of the adoption has barely reached 1.6% of the total houses across the country (GAS 2019b), which indicates an existing problem that needs to be comprehensively investigated.

1.2 Previous Literature

While a tremendous number of researches has been conducted on sustainability transition and human interaction with RSPSs in different contexts, the number of studies conducted in the Saudi context in this regard has been found to be extremely limited. This can be attributed to the fact that individual households have only guaranteed the right to integrate their RSPSs in 2018. As the research and development of solar energy in Saudi Arabia have started a long time before, most of the available literature is technical-focused. A summary of the key literature is provided below:

A study conducted by AlYahya and Irfan (2016a) discussed both the technical and economic opportunities for establishing a new solar industry in Saudi Arabia. The most likely growth scenario of electricity in Saudi Arabia between 2010 and 2025, is the escalation of the peak load from 44.945 MW in 2015 up to 61.279 MW in 2025. This growth is coupled with the major solar potential for energy, as AlYahya and Irfan concluded that after analyzing the Renewable Resource Monitoring and Mapping data. A comparison of direct normal irradiance can be an indicator of solar potential, and in January and July 2013 the range of consumption along the Eastern and Western coast was 3.5–7.0 kilowatt hours per square meter per day ($kWh/m^2/day$), while the Central region of Riyadh received 6.0–6.5 $kWh/m^2/day$, and the Northern region received 9.0–9.5 $kWh/m^2/day$. In comparison to other regions in the world, these figures confirm the high potential for developing a promising solar energy sector in Saudi Arabia (AlYahya and Irfan 2016a, b).

For instance, by comparing Saudi Arabia's possible annual yield to two other cases in Germany and California, Saudi Arabia's possible annual yield was recorded at 1400–2100 kilowatt hours/kilowatt peak (kWh/kWp). While in Germany and California, the possible annual yields recorded were only 825–1100 (kWh/kWp) and 1400–1800 (kWh/kWp), respectively (AlYahya and Irfan 2016a, b). Further, the cost of electricity generation from solar energy in Germany and California is 0.44 (€/kWh) and 0.22 (€/kWh), respectively, while it is less than 0.20 (€/kWh) in Saudi Arabia. This study concluded that the enormous increase in electricity needed in the future, coupled with the low cost and high direct normal irradiance found in Saudi Arabia, presents an enormous opportunity for the solar energy industry in this region (AlYahya and Irfan 2016a, b).

Another study by Mosly and Makki (2018) investigated the current status and the willingness to adopt renewable energy technologies by individuals in the Western region of Saudi Arabia. The study used a quantitative approach by collecting cross-sectional data from 416 respondents through a paper and an online questionnaire survey. The level of the respondents' background knowledge of six different renewable energy sources was assessed, including solar energy, wind energy, sea wave, hydroelectric, geothermal, and biomass technologies. The level of background knowledge was evaluated on a scale with two options: whether (I have any background knowledge, or not). The study also ranked the respondents' willingness to adopt renewable energy technologies (RETs) against five perspectives including saving the environment, owning RET, changing the traditional source of energy to RET source, investing in RET-sourced infrastructure, and paying extra for energy by using RET sources instead of traditional sources. These five perspectives were assessed on a scale of three options, which included willing, neutral, or unwilling. The study revealed that participants tended to have more background understanding of RETs in the solar context and less background knowledge of other types of RETs, such as biomass. The order of highest background knowledge to lowest, based on the six RETs investigated in the questionnaires, was shown to be solar, wind, sea wave, hydroelectric, geothermal, and then biomass. This study also identified that economic factors were the primary influencing determinant of the willingness to adopt RETs by the participants. The results also indicated that a participant's educational level was the most significant sociodemographic factor, making it of utmost importance to increasing the awareness level of RETs, facilitating favorable opinions, and ultimately, the adoption of RETs. Further, the sociodemographic elements of employment status and age were also deemed significant determinants of the respondents' willingness to save the environment and pay extra to use RET sources (Mosly and Makki 2018).

Alrashed and Asif investigated the end-users' perception of adopting different types of renewable energy technologies (Alrashed and Asif 2015). A web-based survey and in-person approaches were used to collect 453 responses covering 16 major cities in Saudi Arabia. Their study aimed to understand the most preferred renewable energy technology among individuals that might adopt such technology; technologies included were solar PV systems, solar water heaters, and micro-wind turbines. They claimed that the participants preferred solar PV systems, and the majority of the participants showed a more positive attitude to purchase electricity generated by renewable sources rather than by conventional sources, even if the tariff was slightly more expensive. Although Alrashed and Asif did not investigate in detail what possible factors might be behind the respondents' answers, the significantly high preference rate toward solar energy may indicate promising potential development of solar PV system adoption by the residential sector in Saudi Arabia.

Furthermore, (Dilek et al. 2015) examined the public perceptions of green energy and willingness to use RSPSs by individuals who live in Dammam city, the capital of the Eastern Provision of Saudi Arabia. By using both hard copies and online questionnaires, a total of 100 responses were collected. The results of assessing the participants' general understanding of renewable energy reveal a lack of some essential knowledge about these types of energy. While 83% of the participants identified solar energy as renewable, only 49% and 26% identified wind and biomass as renewable resources of power. Surprisingly, 18% of the study sample think of oil as a renewable resource of energy, while 21% of the participants see nuclear as a renewable form of power.

Moreover, 85% of the respondents see a need to shift from fossil fuels to use other resources and 88% consider fossil fuel as the leading cause of global warming. Results also revealed that 47 and 35% agree or strongly agree using RSPSs. While 74% of the participants were motivated by the potential reduction of their electricity bills, 66% were open to using RSPSs driven by the desire to fight against global warming phenomena. Moreover, only 73% of those who accepted using RSPSs were willing to pay extra money incurred in the installation and maintenance of RSPSs. While the proportion of acceptors was found to be high, only around 28% were willing to pay less than SAR 20,000 (USD 5,330), and a similar proportion was willing to pay between SAR 20,000–40.000 (USD 5,330–10,660). Nevertheless, the results suggest that a provision of governmental financial support would encourage 87% to consider installing these systems in their residences. It is worth mentioning that is study was reported in early 2015 three years prior to the introduction of the Net Metering scheme which entered into force in July 2018.

In another similar study conducted by Dilek's group (Dilek et al. 2018), the disposition of the individuals in the Eastern Province of Saudi Arabia towards a smarter electricity grid was investigated. In-Person interviews with 228 individuals were carried out analyzing their outlook toward smart electric grid and the use of solar PV, biomass, oil nuclear and wind energies. The results showed that the vast majority of

75% identified oil as the main source; however, 25% indicated other options. When the participants were asked to identify the renewable energy options on the list, only 3% correctly identified wind, solar, and biomass as renewable energy sources. Moreover, 35% omitted biomass, while 32% of the respondents omitted both biomass and wind, and as high as 28% did not choose any of the correct answers. Surprisingly, 16% of the participants perceive oil as a renewable resource. These results show a deficient level of knowledge about energy forms among the residents of the area covered by this study, which interestingly known as a Province that encompasses several industrial facilities and petrochemical plants. However, when the participants were asked about the impact of fossil fuel on the environment, 77% were aware of its damaging effect. Although the work of the Dilek's group has been published, personal contact with Dilek was made to inquire about the sampling method used to collect their study population. Dilek replied: "*We trained our own undergraduate students on how to conduct a face-to-face interview, and they interviewed their household members (sometimes father, sometimes husband)*" (Dilek 2019). Besides the small size of the survey population, the sampling method used in this study might arguably increase the potential for biased results.

A study based on a descriptive analysis carried out by Dhawi (2018) assessed Saudi citizens' awareness about solar energy represented by photovoltaic systems. Through a web-based questionnaire, a total of 424 respondents completed 13 questions regarding the tendency of installing RSPSs and the suitability and confidence in this technology. Dhawi suggested that around 75% of the participants believe that the benefits they can get from solar energy are greater than those of other renewable energy sources, such as wind or earth energy. Moreover, solar energy, wind energy, and earth energy are seen as alternative natural energy sources of electricity by 35%, 8%, and 3% of the participants respectively. While the remaining portion of the study's participants (54%) see in all these energy forms alternative natural sources that can be used for electricity generation. The results also showed that around 30% of the participants believe that solar energy is the sole source of power where they reside.

On a slightly different level, a study conducted by Susilawati and Al Surf (2011) reported on an exploratory survey investigating the level of public awareness towards sustainable housing in Saudi Arabia. The survey was deployed online, collecting 693 responses from members of the Saudi Council of Engineers, which is a professional body that gathers engineers from all fields. Around 90% of the survey population were males, and 77% of the participants were between 21–40 years old. Among the findings of their study, Susilawati and Al Surf reported that more than 70 of respondents agreed that sustainable housing contributes to reducing energy bills, and 71% were open to retrofit their dwellings be sustainable when possible. While on the other hand, 60% agreed to shift from non-solar heating water tanks to a solar heating system. Although these findings may sound encouraging, it is arguably difficult to generalize them to the Saudi population. This is due to the fact that all the participants in this study were members of the Saudi Council of Engineers, who expected to be more knowledgeable and have more interest in implementing environment-friendly technologies. Table 1.1 summarizes the previous key studies

Table 1.1 A summary of previous key studies on the interaction of individuals with renewable energy

References	Title	Methodology		Sample Size	Technology type							Targeted area					Key features
		In-person	Web-based		Solar PV	Biodiesel /biogas	Solar water heaters	Sea wave	Hydroelectric	Geothermal	Micro-wind turbines	Central	Eastern	Southern	Western	Northern	
(Alrashed and Asif 2015)	An Exploratory of Residents' Views Towards Applying Renewable Energy Systems in Saudi Dwellings	●	●	$n =$ 453	●		●				●	●	●	●	●	●	Exploring public views of various types of renewable energy technologies
(Mosly and Makki 2018)	Current Status and Willingness to Adopt Renewable Energy Technologies in Saudi Arabia	●	●	$n =$ 416	●	●		●	●	●	●						Investigating the current status and the willingness to adopt
(Dilek et al. 2015)	Public Acceptance of Renewable Energy and Smart-Grid in Saudi Arabia	●		$n =$ 100	●	●					●		●				Measuring the public's knowledge regarding renewable energy sources, their willingness to use solar energy as the main source in their residence units

(continued)

Table 1.1 (continued)

References	Title	Methodology		Sample Size	Technology type							Targeted area					Key features
		In-person	Web-based		Solar PV	Biodiesel /biogas	Solar water heaters	Sea wave	Hydroelectric	Geothermal	Micro-wind turbines	Central	Eastern	Southern	Western	Northern	
(Dilek et al. 2018)	A smarter Electricity Grid for the Eastern Province of Saudi Arabia: Perceptions and Policy Implications	●		$n =$ 228	●	●					●		●				Assessing an individual's knowledge about green energy sources and their environmental impacts and public disposition toward a smarter electric grid
(Susilawati and Al Surf 2011)	Challenges Facing Sustainable Housing in Saudi Arabia: A Current Study Showing the Level of Public Awareness		●	$n =$ 693	●		●					●	●	●	●	●	Exploring challenges facing sustainable housing in Saudi Arabia by analyzing quantitative and qualitative data collected from members of the Saudi Council of Engineers

(continued)

Table 1.1 (continued)

References	Title	Methodology		Sample Size	Technology type							Targeted area					Key features
		In-person	Web-based		Solar PV	Biodiesel /biogas	Solar water heaters	Sea wave	Hydroelectric	Geothermal	Micro-wind turbines	Central	Eastern	Southern	Western	Northern	
(Dhawi 2018)	Renewable Resources Future and Awareness in Saudi Arabia, Solar Photovoltaic as A Model		●	$n =$ 424	●					●	●	●	●	●	●	●	Evaluating the citizens' acceptance and awareness about the use of renewable energy use through 13 questions deployed online

Source own compilation

on individuals' interactions with renewable energy in Saudi Arabia.

Societal transitions have gained much attention in the past few years. Many studies investigated how innovation penetrates social systems, and related actors interact between and with various layers on socio-technical systems. Such studies have helped to understand the process of innovation development, elucidating factors that cause resistance to dissemination, which often varies based on the context of the case being studied.

Unlike the western context, societal transition studies in the Arab region in general and the hydrocarbon-rich GCC states, in particular, have not been widely explored. Several theories have been proposed to analyze technology transitions in social systems. One of these models is the *multi-level perspective* (*MLP*). One of the substantial studies conducted in the Arab context using this tool was by Kumetat (2012). Kumetat applied the *MLP* in his Ph.D. book as an analytical tool and further developed it for the analysis of the hydrocarbon-rich Arab states, conducting a comparative study between two cases, the Democratic Republic of Algeria and the United Arab Emirates (UAE). Further details on Kumetat's work will be provided in Chap. 5.

A key study conducted in the Saudi context is by Al-Saleh (2011) who studied the potential emerging of sustainable innovation systems in renewable energy systems. In his work, Al-Saleh suggested an empirical investigation, guided by the functions approach that was borrowed from other innovation studies. Though conducting 15 qualitative interviews with stakeholders in Saudi Arabia gained insights into the nature of renewable energy. By looking at seven key functions (entrepreneurial activities, knowledge development (learning), knowledge diffusion through networks, guidance for the search, market formation, resource mobilization and creation of legitimacy), Al-Saleh provided insights into what he argued as barriers impeding the development of renewable energy in Saudi Arabia.

1.3 Previous Research Gaps

This subsection aims to acknowledge that gap in the literature, to help develops research questions and framework that are expected to contribute to the existing literature on transitions toward sustainability (in this case, the transition to RSPSs).

The review of the literature reveals that there several gaps that need to be addressed to understand the current prosses of the RSPSs development in Saudi Arabia. Key research gaps can be described as below:

The available literature's lack of comprehensiveness, when it comes to investigating social perceptions, does not propose concrete levels of analysis regarding potential motives or obstacles to the adoption of RSPSs. Most of the questions that appeared in the literature conducted in the Saudi context might provide a basic level of knowledge on how one type of renewable energy is more preferred than the others. However, it is doubtful such questionnaires can provide saturated data upon which policies can be designed.

Furthermore, as the impact of global warming become more severe and education level on such matter advances, literature shows that more and more individuals tend to become in favor of using renewable energy sources over other environment-harming sources. The key factor herein is to understand to what extent these renewable energies are accepted precisely. The coloration between the respondents' perception of protecting the environment and the willingness to pay extra money to help to address this matter has not been fully explored in the previous studies conducted in the Saudi context. For instance, in the (Mosly and Makki 2018)'s study, they did not precisely investigate to what extent the respondents in Saudi Arabia are willing to pay extra money to use RETs and to protect the environment. They rather set three options to answer these questions (i.e., Willing, Neutral, Unwilling). As demonstrated by previous literature conducted in other countries, these are critical areas that can provide valuable insights into explaining the public's perception in more clearly if investigated in more detail. This indicates the need to understand more precisely how Saudi residents perceive this technology, and before that, understand how they perceive fossil fuel as the main current source of power in the country.

Additionally, the sample sizes covered in the literature are arguably too small to draw solid conclusions. With such a small number of respondents, it might be difficult to argue the appropriateness of a representative sample of a certain population. Moreover, the majority of the literature tended to focus on one professional society (Susilawati and Al Surf 2011) or one of the five main areas in Saudi Arabia (Dilek et al. 2018, 2015; Mosly and Makki 2018) where the results may not be sufficient to generalize the findings to other areas.

Moreover, some of the studies investigating the social preference of RSPSs were conducted before the Net Metering scheme came to force in July 2018, when individuals were allowed for the first time to integrate their RSPSs to the public electric grid. Furthermore, to our knowledge, the available literature that analyzed the public views, did not include analyzing stakeholders from other levels, (such as government stakeholders, local developers, etc.).

On the other hand, studies on sustainability transition in Saudi Arabia are found to be particularly limited. Although (Al-Saleh 2011) study is considered an important reference, it is important to mention that al-Saleh assessed the situation of renewable energy in his study between 2008–2011, back where the renewable energy sector did not exist. Thus, most of his results are outdated and may not precisely explain the current situation. In addition, al-Saleh's study was limited by interviewing stakeholder elites at the intuitional level and did not incorporate an analysis of the individuals' perceptive. Considering the nature of the case studied, we argue that the key excising gaps in understanding the current transition of RSPSs in Saudi Arabia are at the convergence between three main areas; technological transitions, renewable energy, and social perception (Fig. 1.4).

Finally, while the appropriate design of policies for RSPSs can significantly contribute to the successful dissemination of this technology, RSPSs is at the same time is directly related to individual households were their perception is a key element for accepting or rejecting this technology. Thus, an interdisciplinary multi-faceted study that comprises these aspects is needed.

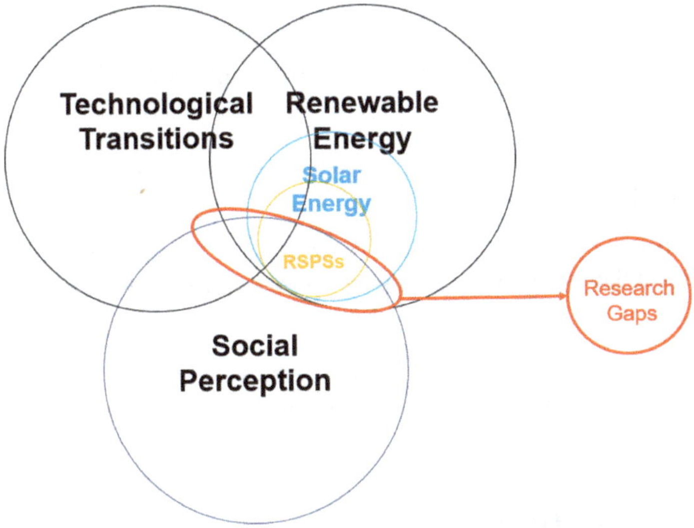

Fig. 1.4 Research gaps and directions

1.4 Research Objectives, Questions, and Hypotheses

1.4.1 Research Objectives

The overall purpose of this book is to understand the current situation regarding the development of RSPSs in Saudi Arabia through achieving the following objectives:

1. To elucidate, from Saudi residents' perspective, the level of acceptance or rejection of RSPSs.
2. To investigate the most critical factors that act as determinants of the individuals' decision to adopt RSPSs.
3. To analyze the factors that shape the policies of RSPSs in Saudi Arabia.

1.4.2 Research Questions

In order to achieve the objectives of this study, the following main research questions have been developed and presented by each chapter.

Chapter 3

(RQ1) How individuals in Saudi Arabia perceive energy options?

Chapter 4

(RQ2) To what extent do individuals in Saudi Arabia accept the notion of installing a residential solar PV system?

(RQ3) What are the most critical factors influencing the individuals' decision to accept or reject the adoption of a residential solar PV system in Saudi Arabia?

Chapter 5

(RQ4) How should the energy policies of the residential solar PV systems and the transition towards it in Saudi Arabia be understood?

1.4.3 Research Hypotheses

This book will attempt to test the validity of the following postulated hypotheses which are presented separately by each chapter.

Chapter 3

(H1) There is a specific preference among the Saudi population against using crude oil.

(H2) The majority of the Saudi population is willing to install RSPS if they would receive subsidies equivalent to 40% of the total cost.

Chapter 4

(H3) The majority of the Saudi population accepts the notion of installing a residential solar PV system in their homes.

(H4) The most significant deciding factor in accepting residential solar PV systems in Saudi Arabia is the *revenue* obtained by reducing electricity consumption and selling excess electricity to the service provider.

Chapter 5

(H5) There are no identified goals of RSPS planned capacity or how it should be implemented in alignment with the national renewable energy plans.

(H6) The participation of the independent Civil Society Organizations in the formulation of the RSPSs policies is significantly limited, with a neglect of the soft-power elements, such as religion and country branding as potential promotion tools.

1.5 Research Structure, Research Interconnected Relationships, and Research Originality

1.5.1 Research Structure

This book comprised six chapters. The content of each chapter is presented below, followed by a presentation of the research framework.

Chapter 1. Introduction

In this chapter, the background, objectives and, motivations of this research are presented. The early solar energy development and rationale for the research are also presented. Further, the previous literature and the literature gaps are identified in this chapter. Lastly, the research questions, hypotheses, and the originality of this research are also stated.

Chapter 2 Global Practices: Assessing Renewable Energy Policies

In this chapter, a review of international experience in disseminating renewable energy technologies is conducted as an entry point for this research. With a particular focus on solar energy, and besides reviewing recent development of the case of Saudi Arabia, the most effective and ineffective practices of seven leading countries, including Germany, Japan, the USA, the UK, Spain, India, and Greece, between the period of 1998–2016, were summarized. Further, the most ineffective actions made by governments of each respective country were categorized into four categorizations (policy, technology, finance, management) based on their main cause.

Chapter 3 An Exploratory Study of the Public's Views on Residential Solar Photovoltaic Systems in Hydrocarbon-rich Saudi Arabia

In this chapter, an empirical exploratory investigation of the individual households' perception of environmental situation and energy options was conducted. Using a non-probability convenience sampling, a cross-sectional web-based survey was employed covering the five main areas in Saudi Arabia (Central, Northern, Eastern, Southern, and Western areas). The questionnaire is divided into two parts; the first part is presented in this chapter, while the other is presented in Chap. 4. The first part of the questionnaire, presented in this chapter, mainly focuses on five aspects, including basic demographic characteristics, residential status, energy consumption, environmental and energy cognition, and willingness to adopt RSPSs.

Chapter 4 Critical Factors Influencing Social Perception of Residential Solar Photovoltaic Systems in the Hydrocarbon-rich Saudi Arabia

In this chapter, a study is carried out in a descriptive style to get an in-depth understanding of influencing determinants on the adoption of RSPSs in Saudi Arabia. The second part of the questionnaire used in this study comprises three sections, including the demographic status section, solar energy cognition section, and an innovation attributes assessment section. Using the *diffusion of innovations* (*DOI*) theory by Rogers (1983) as a framework for analysis, critical factors of accepting or rejecting the notion of adopting RSPSs are assessed on the Rogers' innovation attributes (*relative advantage, observability, complexity, compatibility*).

Chapter 5 Multi-level Perspective of Technological Transitions: Analysis of the Residential Solar Photovoltaic Systems Policies in Hydrocarbon-rich Saudi Arabia

In this chapter, the *MLP* governance approach was used as an analytical tool to capture the socio-technical transition of RSPSs by integrating different explanatory elements in various dimensions. Based on the three integrated layers of the *MLP* model [Landscape-Level (Macro), Regimes-Level (Meso), and Niches-Level (Micro)], a questionnaire was developed and used by conducting 32 face-to-face semi-structured interviews with stakeholders of the renewable energy sector in Saudi Arabia. The outcomes of the semi-structured interviews were analyzed manually by identifying both similarities and differences among the answers of the interviewees. The underlying goal of these interviews is to gather information and knowledgeable opinions on the current transition in the solar energy landscape in Saudi Arabia, preparing the ground for an integrated Transition Management *(TM)* policy design model to develop solar energy policies for the RSPSs case.

Chapter 6 Conclusion

This chapter presents the results of the previous chapters where they are summarized under each research question. Next, an evaluation of the study hypotheses is present. Furthermore, the overall conclusions and remarks are given. Finally, the key contributions and limitations of the research are also highlighted, and recommendations are suggested (Fig. 1.5).

1.5.2 Research Interconnected Relationships

The hierarchy presented below illustrates the interconnected relationships between the elements explored in this study and the respective chapters they represent. Considering the scope of this study, in Chap. 2 the practices of the world's leading countries in solar energy are reviewed and summarized as an entry point to the present study. This is followed by an in-depth multi-faceted analysis of the RSPSs situation in Saudi Arabia. While Chaps. 3 and 4 shed light on the individual households' perception on RSPSs attempting to explore their perspectives and identify the critical factors that influence their decision-making, Chap. 5 attempts to capture the picture of the RSPSs transition and analyzes its policies on a broader level providing lessons learned, and insights on how to guide optimal outcomes. It is hoped that the presented structure will provide an adequate explanation of the status quo of the RSPSs development in Saudi Arabia (Fig. 1.6).

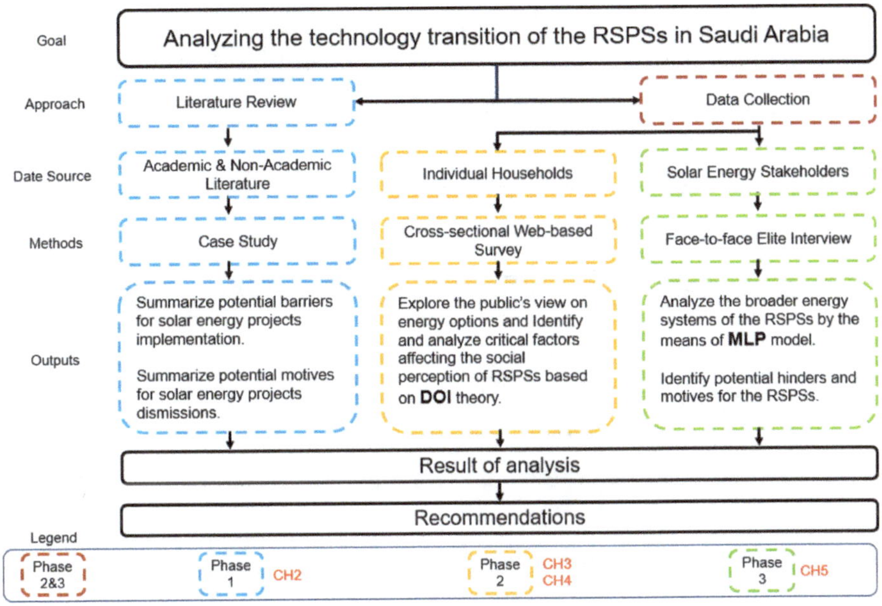

Fig. 1.5 Schematic presentation of research framework

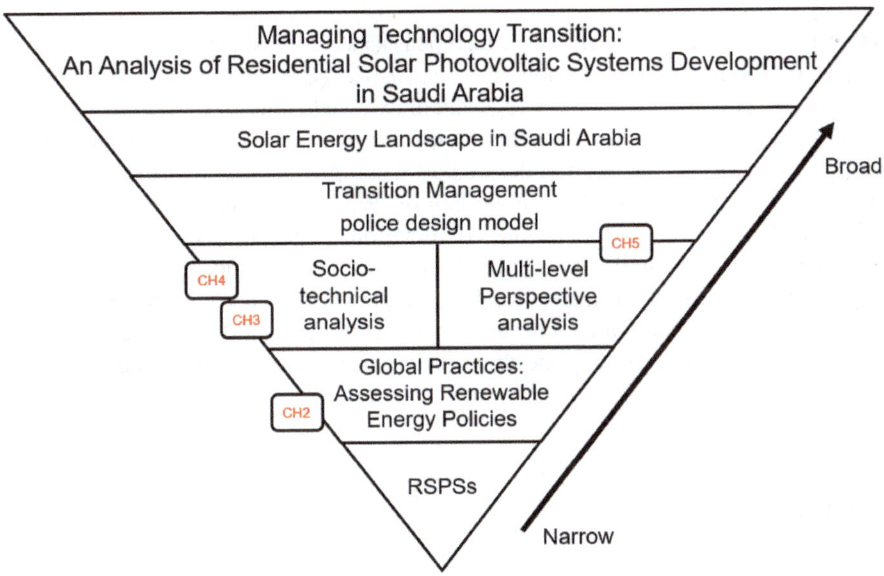

Fig. 1.6 Research interconnected relationships as a hierarchy

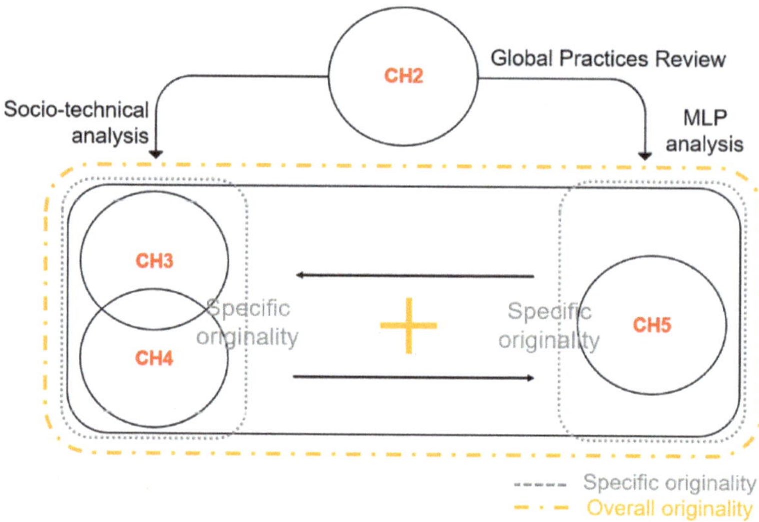

Fig. 1.7 Research specific and overall originality

1.5.3 Research Originality

While many studies focus either on technical aspects, or social aspects, or policy aspects, this book incorporates both the social aspect (Chaps. 3 and 4) and the policy analysis aspect (Chap. 5). The findings and draw conclusions of this study are based on first-hand collected information from the relative stakeholders. The present study has taken several approaches in order to draw a comprehensive picture of the studies case, in a way that has not been done before. While each chapter provides novel information on the case studied, the combination of the three chapters is argued to be the overall originality of this book. This study is considered significant not only because it carried out a large-scale, in-depth analysis among both individuals and institutional levels, but also because it was conducted after the regulations that allowed householders to integrate their RSPSs to the public grid came into force (Fig. 1.7).

References

Alawaji, S. H. (2001). Evaluation of solar energy research and its applications in Saudi Arabia—20 years of experience. *Renewable and Sustainable Energy Reviews, 5,* 59–77. https://doi.org/10.1016/S1364-0321(00)00006-X.

Almasoud, A. H., & Gandayh, H. M. (2015). Future of solar energy in Saudi Arabia. *Journal of King Saud University—Engineering Sciences, 27,* 153–157. https://doi.org/10.1016/j.jksues.2014.03.007.

Alrashed, F., & Asif, M. (2015). An exploratory of residents' views towards applying renewable energy systems in saudi dwellings. *Energy Procedia, 75,* 1341–1347. https://doi.org/10.1016/j.egypro.2015.07.207.

Al-Saleh, Y. M. (2011). An empirical insight into the functionality of emerging sustainable innovation systems: the case of renewable energy in oil-rich Saudi Arabia. *IJTIS, 1,* 302. https://doi.org/10.1504/IJTIS.2011.042662.

AlYahya, S., & Irfan, M. (2016a). The techno-economic potential of Saudi Arabia's solar industry. *Renewable and Sustainable Energy Reviews, 55,* 697–702. https://doi.org/10.1016/j.rser.2015.11.017.

AlYahya, S., & Irfan, M. A. (2016b). Analysis from the new solar radiation Atlas for Saudi Arabia. *Solar Energy, 130,* 116–127. https://doi.org/10.1016/j.solener.2016.01.053.

Chevalier, J., & Geoffron, P. (2016). *The new energy crisis: Climate, economics and geopolitics.* Springer.

Dhawi, F. (2018). Renewable resources future and awareness in Saudi Arabia, solar photovoltaic as a model. *Low Carbon Economy, 9,* 101–108. https://doi.org/10.4236/lce.2018.94008.

Dilek, D. (2019). *Personal communication through E-mails with Dilek Dustegor*, 2019.12.17.

Dilek, D., Nahid, S., Noor, F., & Deemah, A.-Q. (2015). Public acceptance of renewable energy and Smart-Grid in Saudi Arabia | Request PDF, in: ResearchGate. Presented at the Proceedings of the 8th IEEE GCC Conference and Exhibition, Muscat, Oman. https://doi.org/10.1109/IEEEGCC.2015.7060018.

Dilek, D., Sultana, N., Felemban, N., & Al Qahtani, D. (2018). A smarter electricity grid for the Eastern Province of Saudi Arabia: Perceptions and policy implications. *Utilities Policy, 50,* 26–39. https://doi.org/10.1016/j.jup.2017.09.009.

EIA [WWW Document]. (2016). overview india, Analysis, U.S. Energy Information Administration (EIA). Retrieved 9, Dec 2019. https://www.eia.gov/beta/international/analysis.php?iso=IND.

GAS. (2019b). 1.60 percent of households using solar energy at residence in 2019 [WWW Document]. General Authority of Statistics. Retrieved 22, Dec 2019, https://www.stats.gov.sa/ar/1060.

Hepbasli, A., & Alsuhaibani, Z. (2011). A key review on present status and future directions of solar energy studies and applications in Saudi Arabia. *Renewable and Sustainable Energy Reviews, 15,* 5021–5050. https://doi.org/10.1016/j.rser.2011.07.052.

IMF. (2016). International Monetary Fund, Economic Diversification in Oil-Exporting Arab Countries. Policy Papers 16. https://doi.org/10.5089/9781498345699.007.

Jalilvand, D. R., & Westphal, K. (2017). *The political and economic challenges of energy in the middle East and North Africa*, 1st ed. Routledge.

Kumetat, D. (2012). Managing the transition: An analysis of renewable energy policies in resource-rich arab states with a comparative focus on the United Arab Emirates and Algeria. The London School of Economics and Political Science, London.

Luciani, G. (2013). *Political economy of energy reform: The clean energy-fossil fuel balance in the Gulf*. Berlin: Gerlach Press.

Mosly, I., & Makki, A. (2018). Current status and willingness to adopt renewable energy technologies in Saudi Arabia. *Sustainability, 10,* 4269. https://doi.org/10.3390/su10114269.

Mouritsen, S. (2018). Black Gold in a changing world: An examination of Saudi Arabia's dependence on oil and the possibility of a solar energy transition. Independent Study Project (ISP) Collection.

Nachet, S., & Aoun, M.-C. (2015). *The Saudi electricity sector: pressing issues and challenges.*

Nash, J., Alyahyai, A., & Koyame-Marsh, R. (2017). Silicon in Saudi Arabia and its potential impact on the Saudi economy. *International Journal of Mining and Mineral Engineering, 8,* 144. https://doi.org/10.1504/IJMME.2017.084205.

Rogers, E. M. (1983). *Diffusion of innovations*, 3rd Edition, 3rd ed. Free Press ; Collier Macmillan, New York : London.

Sakr, I. (1987). Solar energy applications in the Arab world, Proceedings of the International Seminar on Appropriate Technology in the Fields of Solar and Wind Energy Applications.

Saudi Arabia Carbon (CO2) Emissions 1960-2019 [WWW Document], 2019. Retrieved 26, Dec 2019, from https://www.macrotrends.net/countries/SAU/saudi-arabia/carbon-co2-emissions.

Sullivan, P. (2016). *Saudi Arabia's Oil-economy roller coaster & vision 2030.* Washington, D.C., USA: National Council on U.S.-Arab Relations.

Susilawati, C., & Al Surf, M. (2011). Challenges facing sustainable housing in Saudi Arabia : a current study showing the level of public awareness. Presented at the The 17th Pacific Rim Real Estate Society Conference, Bond University, Gold Coast, Qld.

Tagliapietra, S. (2017). The political economy of Middle East and North Africa oil exporters in times of global decarbonisation | Bruegel. Retrieved 26, Dec 2019, from, https://bruegel.org/2017/04/the-political-economy-of-middle-east-and-north-africa-oil-exporters-in-times-of-global-decarbonisation/.

The International Energy Efficiency Scorecard [WWW Document], 2018.. ACEEE. Retrieved 26, Dec 2019, https://aceee.org/portal/national-policy/international-scorecard.

The World Bank. (2000). Fuel for thought: An environmental strategy for the energy sector (No. 20740). The World Bank.

Tlili, I. (2015). Renewable energy in Saudi Arabia: Current status and future potentials. *Environment, Development and Sustainability, 17,* 859–886. https://doi.org/10.1007/s10668-014-9579-9.

Chapter 2
Global Practices: Assessing Renewable Energy Policies

In this chapter, the most effective and ineffective policies and actions implemented by seven selected countries to encourage the deployment of solar power will be discussed and summarized. Additionally, the recent development of solar energy in Saudi Arabia is also dissected. The key actions that turned out to be ineffective were then categorized into four groups (*policy*, *technology*, *finance*, and *management*), for the sake of a better understanding of the nature of causes behind these actions.

2.1 Definition of Terms

In order to better understand each case of the selected countries, the researcher presents the most significant actions taken by each country that have positively or negatively affected the development and the implementation of solar energy projects. The actions that have positively influenced the implementation of solar energy projects and helped to boost and promote this technology are typically referred to as *effective actions*. On the other hand, activities that may have caused a stumble or delay in the development of solar energy are referred to as *ineffective actions*. These ineffective actions are then classified into four categories for the purpose of better understanding the main causes of the issues which were responsible for hindering the implementation of solar energy projects. First, the term *technology*, which indicates failures caused by the poor performance of solar systems, include solar panel, inverter, battery pack, controller, etc. Whether the poor performance results from internal factors, such as material defeat, or external factors, such as harsh weather, it is categorized under the term of *technology*. *Technology* is defined as "Machinery and devices developed from scientific knowledge" (Oxford Dictionaries 2017a). It is also interpreted as "The application of scientific knowledge for practical purposes, especially in industry" (Oxford Dictionaries 2017a). In this chapter, the term *technology* is used for solar system applications that stumbled due to system failures or

K. A. Alrashoud, *Managing Technology Transition in Saudi Arabia*,
https://doi.org/10.1007/978-981-15-5803-0_2

technicians' lack of training. Second, the term *policy,* which is identified as "A course or principle of action adopted or proposed by an organization or individual" (Oxford Dictionaries 2017b). This covers regulatory mechanisms adopted by a government to organize the relationship among electricity sellers, buyers, and end-users. It also includes policies made to decrease ambiguity and uncertainties for investors, as well as to set new policies aiming at fostering the attitude of the adoption of solar energy.

As for the term of *finance*, the current sense of the term can be traced to at least the eighteenth century. "Late Middle English: from Old French, from finer 'make an end, settle a debt,' from fin 'end.' The original sense was 'payment of a debt, compensation, or ransom'; later 'taxation, revenue'" (Oxford Dictionaries 2017c). In this research, the term finance includes investment risk, financing availability, interest rate risk, inflation impact, and tax regime risk. Finally, *management* is referred to as the technics used to manage and organize the available resources. Oxford English Dictionaries identified it as "The process of dealing with or controlling things or people" (Oxford Dictionaries 2017d). Whether it is a governmental body, a corporation, or a non-profit organization, in this book, the term *management* indicates the action of administrating the respective organization's resources. It also includes the activities of coordinating the efforts of the organization's employees to achieve its objectives within the available resources, such as financial, technological, and human resources. The term *management* may also refer to managing the end-users who make use of solar energy applications provided by government agencies or private-owned firms in the respective country.

2.2 Methodology and Research Design

The case study approach was adopted to conduct this research. "The case study approach allows in-depth, multi-faceted explorations of complex issues in their real-life settings" (Crowe et al. 2011). "Case study research, through reports of past studies, allows the exploration and understanding of complex issues" (Zainal 2007). It is considered as a robust research method, particularly when a holistic investigation is required. Besides Saudi Arabia, seven countries (Germany, Japan, the USA, the UK, Spain, India, and Greece were selected based on their rich experience in creating new markets for renewable energy technology and confronting plenty of challenges and difficulties along the way. The date was gathered from the available literature, magazines, reports, and newspaper articles. With a special focus on solar energy, the main effective and ineffective renewable energy-related actions made by each country between 1998–2016 were summarized and categorized into four groups (*Policy, Technology, Finance, and Management*). However, only the recent development of solar energy was covered in the case of Saudi Arabia, particularly, the period of 2016–2019. Since the present study mainly focuses on countries that rely on their local developers and individual households to spread the use of RE, countries with massive government-driven development such as China were excluded from this

Table 2.1 Power generation by source in Saudi Arabia, 2016

Saudi power generation share by fuel (%)	
Natural gas	50.7
Crude oil	24.2
Heavy fuel	16.5
Diesel	8.6
Renewables (solar)	0.04

Source (Krane 2019)

research. This is to specifically highlight how the governments of the seven respective countries interlinked with local developers and end-users and to identify difficulties confronted by each county when creating new markets for renewable energy.

2.3 Results and Discussion

2.3.1 *Renewable Energy Policies in Saudi Arabia*

Despite its exceptional solar energy potential, Saudi Arabia produced over 99% of its needed electricity in 2017 from fossil fuels (Table 2.1). In the same year, the Kingdom has generated 376 terawatts (TW) through its 88 (GW) generating capacity. A mere 0.04% of that energy produced from solar energy. Over the past few decades, the Saudi energy sector has evolved rapidly in terms of power generation, which has been increasing by an average of 6.4% annually since 2015 (Krane 2019). Where oil-based liquid fuels once covered over half of the needs of feedstock, the expansion of the country's natural gas production and installation of gas-fired generation by 2016 gave natural gas a slight advantage in terms of generated power, along with a 60% share of installed capacity (Krane 2019).

After several failed attempts to develop large-scale renewable energy projects under the K.A.CARE initiatives, the Saudi government established the Renewable Energy Project Development Office (REPDO), an office within the Ministry of Energy[1] to deliver the 9.5 GW renewable energy capacity by 2023 set by Saudi Vision 2030 (REPDO 2017). In early 2019, this target was amended and increased to become a 27.3 GW by 2023 (Fig. 2.1). These projects will be implemented under the umbrella of the "King Salman Renewable Energy Initiative," which is a new

[1]Within the last three years the energy industry has gone under a major restructuring process. The Ministry of Energy, for instance, which was until May 2016 used to be called as the Ministry of Petroleum and Mineral Resources was renamed as the Ministry of Energy, Industry and Mineral Resources with an aim to ensure the alignment of efforts and goals towards further economic growth. Ambitious plans were set for the industry and mining sectors, yet only limited development has been achieved. In August 2019, the Ministry of Petroleum and Mineral Resources was divided into two- Ministry of Energy and Ministry of Industry and Mineral Resources (SPA 2019; TNS 2019).

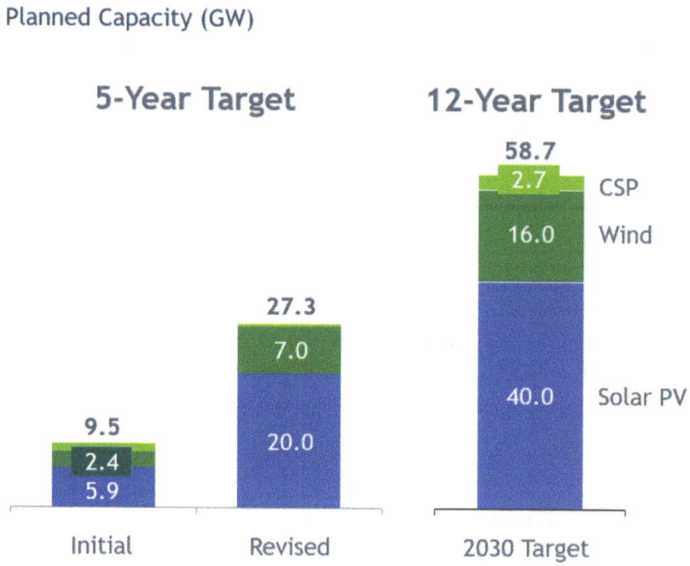

Fig. 2.1 Saudi national renewable energy revised targets

part of the wide-ranging economic and social reforming plan, Vision 2030. These projects will spread across the country to promote regional development and mitigate technology risk.

The revised targets are considered a substantial increase over the previous once, contributing to attracting private investments and enabling price discovery. While 70% of this capacity will be awarded to the country's sovereign wealth fund named the Public Investment Fund (PIF) to negotiate deals with international partners, the development approach for the remaining 30% will be through competitive tenders carried out by REPDO (Turki Alshehri 2019). All projects tendered in 2019 by REPDO were issued under the independent power producer (IPP) arrangement and must be backed by 20 to 25-year power purchase agreements with a minimum requirement of local content. All sites shall go under pre-development stage, which includes activities reduce risk and lower the cost of the Levelized cost of energy (LCOE). These activities include energy yield assessment, grid impact studies, measure resource, environmental assessment, Geotech assessment, hydrological assessment.

Key incentives provided by the Saudi entities for utility-scale project developers can be summarized as follow (Table 2.2).

Since these incentives have been offered very recently, they need to be followed closely in order to further assess the impacts on the development of utility-scale projects. However, these utility-scale projects continue to be developed, in not clear how the distributed RSPSs will be implemented in alignment with these projects.

In parallel to the utility-scale solar project implementation, the Saudi government counts on the small-scale distributed solar systems to balance out its energy portfolio. Although the utility-scale projects are less complicated and obviously can bring

Table 2.2 Incentives offered by Saudi government entities for utility-scale project developers

Incentives	Entity	Description
Duty Exemption	MOE, Saudi Customs	Exempts manufacturers with an industrial license from the duty on raw material and finished products not produced domestically
Low Land	Royal Commission for Jubail and Yanbu, Modon	Provides manufacturers with plot spaces with the required basic infrastructure for a low price of 0.8 $cent/m^2
Favorable Loan Terms	Saudi Industrial Development Fund	Offers 12-year loan financing at up to 75% of capital expenditure with upfront fees of 6–8% & semi-annual fees of 1.5–2.5% of outstanding amounts
Training and Salary Grants	Human Resources Development Fund	For Saudi hires

Sources (Turki Alshehri 2019)

quicker results, what makes the Saudi government pursue the transition through RSPSs Given the temporal and spatial conditions, several reasons can be given. Unlike most developed and developing countries, Saudi Arabia is one of few countries that escaped colonial domination. It has been recognized as a sovereign nation-state since 1932, when King Abdulaziz Al-Saud, the father of the current King Salman bin Abdulaziz, merged tribal lands and placed them under his control. A few years later, a major discovery of oil wells took place, an event that has been a fundamental element in the stability, development, and prosperity of Saudi Arabia. It has also formed the country's political economy and facilitated what has been referred to as the "ruling bargain." The ruling bargain is known as an implicit top-down social contact where governments of the GCC monarchies, grant their citizens wide-ranging welfare benefits in return to their full political support (Mehran 2013). Over the past decades, Saudi Arabia has spent a substantial amount of oil export revenues to maintain this unique citizen-state relationship. Among GCC countries, Saudi Arabia ranked first in energy subsidies as a percent of GDP (Hertog 2015). The escalating spending on subsidies rang the alarm bells of the Saudi economy to take action to avoid an unpleasant future (Fig. 2.2).

Additionally, low-priced energy has created an illusion of abundance, resulting in extremely wasteful consumption habits. These consumption habits have been putting even further pressure on the government to meet the rising demand for both stable energy supply and fulfillment of reducing Carbon dioxide (CO_2) emissions. The government realized that limiting the integration of solar energy to the utility-scale solar parks will not address the issue of the individuals' excessive consumption of electricity.

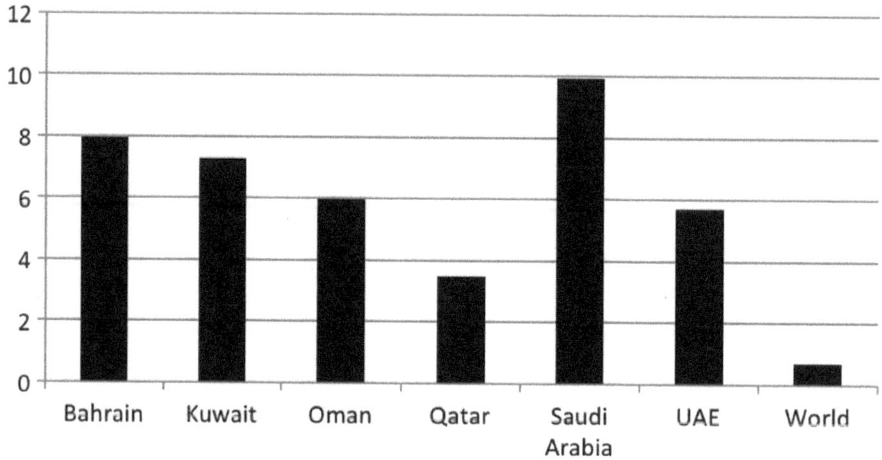

Fig. 2.2 Estimated GCC countries energy subsidies as a percent of GDP

Furthermore, the uprising of the Arab spring pushed by growing populations and high unemployment rates has been a major influencing factor on the Arab youth and their relationship with their governments. Although no notable movements driven by this event have been observed in Saudi Arabia, it encouraged the Saudi government, with a sense of urgency, to take serious steps seeking reforms of its "rentier economy." Since 2016, the Saudi government has been moving towards involving individuals to play an active role in maximizing their contribution to the country's GDP and reduce the pressure from the government's shoulders. Different approaches have been taken to achieve this goal, including imposing taxes, cutting government subsidies, as well as, involving individuals in generating the power they need, a task that had always been the responsibility of the government.

In addition to the reasons mentioned above, Saudi Arabia seeks to create a new sector of a solar energy market and becomes a leading trading hub for the Middle East, rather than merely seeking to import solar technologies and executing mega-scale solar projects through transregional developers. This also goes in line with the country's goals of transferring the related knowledge and localizing these technologies to create new jobs and take advantage of the abundant stock of pure silica sand, which is an essential component in manufacturing PV panels (Saudi Vision 2016b).

The Electricity & Co-Generation Regulatory Authority (ECRA) that regulates the electricity and water desalination industry in Saudi Arabia prepared a regulatory framework for individuals distributed small-scale solar generation units to the national grid. The regulations of this framework entered into force in July 2018 (ECRA 2017). Before to that, the local electric company, the Saudi Electricity Company (SEC), which enjoys a monopoly on the generation, transmission, and distribution of electricity across Saudi Arabia, has called upon individuals for initial applications from September 2017 to gauge how many consumers are interested in installing systems and allow it to predict the impact of the electricity produced on

Table 2.3 Scope and application of the Small-scale solar PV systems regulations introduced by ECRA

No.	Scope and application of the small-scale solar PV systems regulations
1	These regulations shall apply to the distribution service provider (DSP), eligible consumer, certified consultant/contractor, and any other persons involved in the connection of small-scale solar PV systems to the distribution system and/or entering into a Net Metering arrangement with the DSP
2	The eligible consumer may install the small-scale solar PV system under Net Metering arrangement which: (a) shall be within the permissible rated capacity which the maximum connected capacity of small-scale solar PV system to be installed at any eligible consumer's premises shall not exceed the connected load for the consumption account; (b) shall be located in the premises of the eligible consumer; (c) shall not exceed a capacity of 2 megawatts (MW) in one premises; (d) shall not exceed an aggregate capacity of 5 MW installed in different premises owned by the same eligible consumer in the area of supply at one electricity department; (e) shall not be less than 1 kW; and f) shall connect and operate safely with the distribution system
3	These regulations do not apply to solar PV systems greater than 2 MW capacity or smaller than 1 kW capacity or to solar PV systems that do not operate in parallel with the distribution system
4	These regulations do not preclude the right of the investors to undertake solar PV projects of more than 2 MW capacity where there are alternative mechanisms prepared by the principal buyer for such projects

Source (ECRA 2017), own compilation

demand and the public grid. According to ECRA, the scope and application of the regulations are as follows (Table 2.3).

These regulations apply to all categories of electricity consumers (residential, industrial, agricultural, commercial, government), yet, the DSP should prioritize the availability of Net Metering arrangements to residential, commercial, government, and agriculture categories of consumer.

Net Metering is widely used enabling policy designed for electricity consumers to operate their own, small-scale solar power (in case of Saudi Arabia (<2 MW)) generating systems and export surplus electricity (if any) to the public grid, offsetting this volume against their own consumption. The diagram shown below illustrates how the Net Metering billing mechanism operates (Fig. 2.3).

The distributed small-scale connection process, as per ECRA, can be summarized into four steps as shown below in Fig. 2.4. Consumers must follow these steps in order to be able to integrate their RSPSs into the national grid. Through the Net Metering agreement set by ECRA, the consumers are granted the right to export any surplus energy back to the distribution system and recorded in the billing system at the same tariff they purchase it. Surplus energy is carried forward from one billing cycle to the next and shall be offset against future energy consumption at an exit point (ECRA 2017). At present, this scheme considered as the only incentive scheme provided by the government to RSPSs adopters. The current electricity tariffs of all categories of consumptions are shown in Table 2.4.

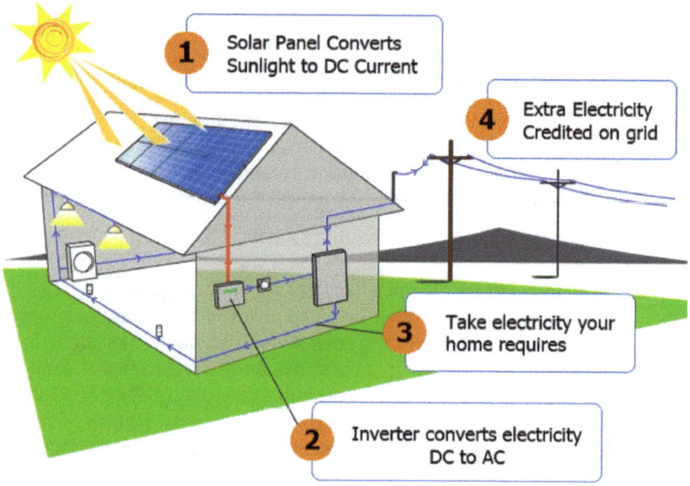

Fig. 2.3 Rooftop Net Metering diagram

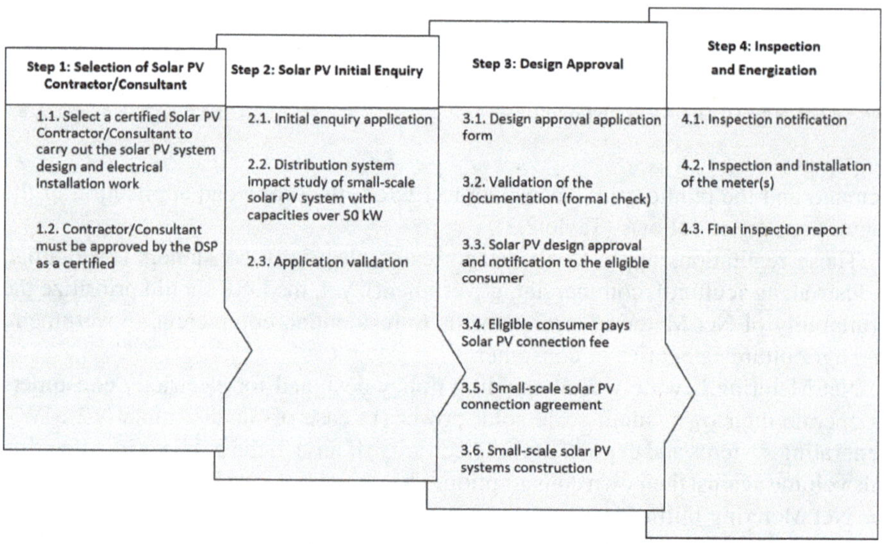

Fig. 2.4 Small-scale solar PV systems connection process as proposed by ECRA

The distributed small-scale connection process, as per ECRA, can be summarized into four steps, as shown below in Fig. 2.4. Consumers must follow these steps in order to be able to integrate their RSPSs into the national grid.

DSPs should set and submit for ECRA's approval a list of fees for connection and inspection to be levied on the consumers. In Table 2.5 the fees vary based on system

Table 2.4 Electricity tariffs for all categories of electricity consumption

Consumption categories (kWh)	Residential (Halalah[a]/kWh)	Commercial (Halalah/kWh)	Agricultural & charities (Halalah/kWh)	Governmental (Halalah/kWh)	Industrial (Halalah/kWh)	Private educational facilities, private medical facilities (Halalah/kWh)
1–6000	18	20	16	32	18	18
More than 6000	30	30	20			

Source (SEC 2018)
[a]SAR = 100 Halalah. 1 Halalah = USD 0.0026

Table 2.5 List of fees for services relevant to the connection of small-scale solar PV systems

Service	System Capacity (kW)	Fees/Charges (SAR)[a]
Initial inquiry application	50 or less	150
	More than 50	500
Small-scale solar PV connection	50 or less	550
	More than 50	1800
Total	50 or less	700
	More than 50	2300

Source (ECRA 2017)
[a]Saudi Riyals (SAR) = USD 0.26

Table 2.6 A summary of actions taken by the government of Saudi Arabia

Country	Total PV installed capacity (MW)	Effective actions taken by government	Ineffective actions taken by government
Saudi Arabia	35.2	Relies on the private sector to develop solar parks and limits the government role to tendering and supervision	Not specifying the nature of components included within the local content obligation
		Conducting pre-development activities on the site to increase certainty and deliver lower project costs	RSPS uptake rate equal to the purchase rate under the Net Metering agreement
		Improving governance and increasing transparency including specifying the capacity of future projects	Absence of tax deduction and soft loans programs
		Obligating successful bidders with a gradual minimum requirement of local content of facilities to be developed	Absence of promotion campaigns and information provision

Source (ECRA 2017)

capacity are presented. Table 2.6 summarizes the Saudi government's actions toward the implementation of solar energy.

2.3.2 *Renewable Energy Policies in Germany*

Germany is one of the major electricity consumers in Europe. In 2013, Germany's electricity production reached 532.2 TWh (EnergyCharts.de 2016). The primary energy source in Germany is brown coal with a capacity of 140.7 TWh (26.4%), followed by hard coal, representing 11.1 TWh (20.7%). The third-largest energy source in Germany is nuclear constituting 91.8 TWh (17.2%), and then biomass comes fourth, with a 53 TWh, representing 10.0%. Next is wind power with 51.4 TWh (9.7%), natural gas 33.9 TWh (6.4%), solar 32.8 TWh (6.2%), and hydro 18.5 TWh (3.5%) (EnergyCharts.de 2016). Regarding RE market development, Germany has been considered as a successful leading model of transition from depending on fossil fuel energy to the use of RES as a clean alternative source of energy. In order to understand the factors behind these achievements, it is necessary to know what it results from.

After the Social Democratic-Green government came into office in 1998, a further boost has been given to the comprehensive promotion plan that launched at the beginning of the 1990s to enhance the utilization of RE. Promotion measures have been enacted to supply 18% of the countries' consumed energy from renewable sources by 2020 (Brown 2013a). The German parliament passed the binding equalization scheme (EEG) legislation that was amended multiple times between 2000 and 2012 to revise and eliminate quotas on certain technologies. By enforcing this legislation and through other policy adjustments, these procedures were necessary to persevere in the rapid expansion of renewable power in Germany, especially with the boom of solar PV applications. Besides, in 2012, the market premium policy for renewable power projects was introduced by the government. This policy indicates that the FiT policy structure will likely shift to other policies, such as the market premium or some other market integration incentives over time. The motivation for introducing such policies was to involve renewable energy projects in the electric power market actively. Hence, premium incentives in Germany are designed to decline gradually over time based on the predetermined and the set market-responsive schedules (Brown 2013b). In 2012, an amendment to Germany's EEG was made to enable renewable electricity projects to receive adjustable market premiums rather than FiT incentives. The EEG policies have many nuances which made it a complex mechanism (Brown 2013c).

The government has also focused on wind and solar PV projects that can be monitored and managed remotely, offering high management fees to incentivize these technologies for more renewable electricity integration. At the same time, the management fee of the market premium has been gradually reduced, starting from January 2013. This gradual reduction was also applied to the FiT. The main goal of the gradual reduction of this fee is to set FiTs at a level that encourages investment in RE projects and expands the market of RE generation while at the same time avoiding nuances of the over-deployment of RE projects, as well as preventing windfall profits for project owners. One approach to stimulate the expansion of RE in Germany is the electricity surcharge, which secures the necessary funds of RE incentives by dividing the costs over a larger electricity consumer base. This differentiates Germany from other countries, as it uses federal tax revenues to fund their incentive programs (Matthias 2013). Table 2.7 summarizes the German government's actions toward the implementation of solar energy.

Table 2.7 A summary of actions taken by the government of Germany

Country	Total PV installed capacity (MW)	Effective actions taken by government	Ineffective actions taken by government
Germany	45.930	Shifting from FiT to tender or quota system by 2017 guarantees grid access and uptakes the surplus power	Supporting unequally split across rare classes made heavy consumers the most levy-exempt
		Setting a cap on aggregate solar PV capacity of 53 GW	Implementing FiT surcharge on consumers to pay the difference between the fixed price of RE fed into the grid and the sale of RE at the energy stock market
		Mandatory direct marketing implementation	Delay in readjusting FiT's rates

2.3.3 *Renewable Energy Policies in Japan*

After the suspension of nuclear reactors due to the Fukushima plant accident in 2011, domestic energy resources in Japan have dropped down from 20% to less than 9% of the country's total energy use. With its massive energy consumption, Japan ranks the third-largest oil consumer after the USA and China in the world. Despite its limited domestic power resources, Japan has actively worked to develop its capability in the energy sector by increasing the use of natural gas and coal, as well as encouraging the use of renewable energy and the efficient consumption and conservation of energy (BP 2016).

Since 2011, the government seeks to cover the shortage of energy supply by shifting to renewable alternatives. Both previous governments, as well as the current government, have implemented several reforms and policies to ensure a stable and safe supply of energy without undermining the majority of Japanese who oppose atomic energy. One of the most significant mechanisms is FiT, which is a key scheme that was introduced in 2012 following the shutdown of the nuclear reactors. It specifically began in July 2012, obliging companies to purchase electricity generated by renewable sources at fixed rates. The revenue received will then be used to facilitate investment in renewable power generation.

In the middle of 2016, the Ministry of Economy, Trade, and Industry (METI) sought a change of the currently existing FiT system (Dokei et al. 2016). The new amendments to the FiT system included a significant number of modifications that set more constraints on the existing rules (Dokei et al. 2016). These constraints include:

- To obligate conducting regular inspections and maintenance to ensure a stable supply
- To submit production incurred costs and install capacity to METI at fixed intervals
- To generate power in a manner that ensures grid stability

Table 2.8 A summary of actions taken by the government of Japan

Country	Total PV installed capacity (MW)	Effective actions taken by government	Ineffective actions taken by government
Japan	55.500	Low-interest loans	Immediate suspension with no previous notification for new projects troubled developers and operators
		Designing comprehensive education and awareness programs across the country	Awarding contracts to some companies without verifying their technological as well as financial abilities
		Introducing Renewable Portfolio Standard (RPS) Law which obliges electric power companies to expand their use of electricity generated from RES	Awarding contracts without specifying a time frame to implement RE projects

- To suspend all facilities that have ceased to be used to link with a utility grid
- To initiate a new supply with a fixed time frame
- To adhere to all land-related laws and safety regulations
- To preserve to signage on-site indicated in the criteria list that must be met by the generation/operation company to be certified.

The new amendments give METI the authority to revoke the infringing plant's certification if one or more of the requirements are not satisfied.

Moreover, it enables METI to oblige project developers/operators to rectify their infringing equipment that has been in use. It has been reported that some of these new constraints doubled up on other legal requirements, resulting in institutional issues. Nevertheless, it is expected that METI will show less tolerance to operators who are not on track to contribute to its objectives of building a robust and diversified electricity supply (Dokei et al. 2016). Table 2.8 presents actions made by the Japanese government regarding the implementation of solar energy projects.

2.3.4 Renewable Energy Policies in the USA

The USA is considered as the world's second-largest energy consumer and producer after China. The primary source of energy in the USA is petroleum, followed by natural gas and coal. It also generates power from clean energy sources, such as biomass, wind, solar, and nuclear power. In terms of renewable energy, the USA has enacted several policies and mechanisms to boost the development of RE to cope with the increasing demand for energy. Renewable energy standard (RES), which is also known as the renewable portfolio standard (RPS), is one of the most successful policies implemented to enhance renewable energy projects in the USA. RPS was combined with federal production tax credit shaping an effective tool to catalyze the utilization of renewable energy technologies. Driven by such combined tools, States have designed them to encourage the adoption of a certain technology (e.g., wind, solar) and provided carve-out provisions to ensure a share of electricity generated from that particular technology. With the flexibility offered by RPS, the States have the freedom to apply the requirement of RPS either to the investor-owned utilities or to all its existing utilities. Moreover, States have the right to determine the eligibility of each type of renewable energy technologies to be counted to fulfill the RPS requirements (NREL 2015).

Solar investment tax credit (ITC) is another key federal policy mechanism established in 2005 to motivate the deployment of renewable electricity projects in the USA. Even though limited supply is seen as a disadvantage, a credit of 30% of qualified expenditures for systems that are installed in a household can be claimed by taxpayers. Another key scheme is renewable energy certificates (RECs) that are designed instruments to track, account, and assign ownership to renewable energy generation and help enhance the development of renewable energy projects based on the market status. Generally, REC is issued when an operator successfully delivers 1 MWh of electricity generated from RES to the power grid (Energy.gov 2016). The Virtual Net Metering (VNM) tool, which is nearly identical in structure to the Standard Net Metering (SNM), was introduced under the California Solar Initiative Multifamily Affordable Solar Housing Program (MASH) as a means of enabling low-income families who are living in apartment complexes to make benefits by receiving credits from the solar system owner (building owner) without being required as low-income tenants to physically connecting their meters to the solar system (SEIA 2015). The Single-family Affordable Solar Homes program (SASH) is also a similar program launched by the same initiative to enable low-income families to own their RSPSs. SASH has also trained nearly 13,000 volunteers in PV systems installation techniques. These two initiatives increased the solar PV capacity in California more than 16-fold (Lindsey et al. 2015).

The Net Metering regulations and compensation policies from a state to another vary based on different factors, such as system size, or technologies. The states have taken a range of different approaches to boost the rooftop solar capacity by addressing excess generation. For instance, in California, excess generation is credited to a customer's next bill at the retail rate. After a 12-month period, customers can choose whether to receive a payment for credits or roll credits over indefinitely (CPUC 2016a; NCSL 2017).

In addition to the mechanisms and policies mentioned above, the USA has been providing financial support in different forms which induced a tremendous growth of integrating PV systems in the dwellings. Programs such as offering repayment guaranteed loans or direct lending of money to energy market participants had a positive impact on expanding the use of solar energy systems. These subsidized loans helped householders with low income to afford the cost of solar systems (CPUC b). On the other hand, recent reports show an increasing system failure which jeopardizes

Table 2.9 A summary of actions taken by the government of the USA

Country	Total PV installed capacity (MW)	Effective actions taken by government	Ineffective actions taken by government
USA	25.620	Federal tax credit	10-year long-term contracts increased insecurity and risk for the generators
		Loan and insurance programs	Weak control caused over reporting
		RPS accessible via the internet	Ambiguity of contract terms to customers
		Locational marginal pricing (Electricity tariffs varies at different locations in the power grid)	Authorization to a monopoly company to charge higher rates at the expense individual households (Nevada's public utility commission)
		Designing programs for low-income families and voluntarily train technicians	

the steady increasing rooftop capacity. According to the National Renewable Energy Laboratory survey, companies attempt to lower the prices of panels by reducing the aluminum and make their frames thinner was a common issue that caused a failure to many installed systems (Jordan and Kurtz 2013). Table 2.9 illustrates the key actions that influenced the introduction of solar energy in the USA.

2.3.5 Renewable Energy Policies in the UK

In aiming to reduce GHG emissions and coping with depletion of the country's natural resources, the UK outlined numerous commitments, including several incentives to expand the use of renewable energy. The primary energy production cost (10% of the UK's GDP) is much higher than most of the other industrialized countries, which put more pressure on the UK to seek other alternatives. In the UK, electricity is generated in many forms. According to 2015 statistics, fossil fuel is the most used source to produce electricity. Natural gas (29.5%) and coal (22%) are the main sources of energy. Other sources, such as nuclear (20.8%) and renewables (27.7%), are major players in the UK's energy mix. Currently, the oil contribution is under 1%. The depletion of its domestic fossil fuel reserves combined with projected growth in global demand urges the UK government to implement plenty of reforming schemes to include renewable sources as a key part of its energy mix (Planète 2015).

Several incentive schemes have been launched in the last decade to stimulate investments of renewable energy for both individuals and businesses. For instance,

Renewable Obligation (RO) was introduced in 2002 to place an obligation on electricity suppliers to receive an increasing share of their electricity from renewable sources. Operators carry out this by purchasing a Renewable Obligation Certificate (ROC), which is issued to all accredited generators for the share of the generated renewable electricity. Therefore, two sources of revenue are available to the generators: earnings from the sale of the ROCs and revenue from selling generated electricity to the wholesale market, regardless of the electricity sourced from renewable or nonrenewable energy. The system of ROC was modified to ensure that all types of renewables are receiving effective support without bias toward a particular technology. Therefore, since 2009, the government has banned issuing one ROC for each MWh generated from renewables as an attempt to create balance and prevent market distortions (Grimwood and Elena 2016). The time-of-use (ToU) in the UK has been found as a significant factor in reducing the overall cost of the electricity system. It also was argued that it is beneficial for consumers as they can take advantage of lower prices under the ToU tariffs (CSE 2013).

Regarding energy efficiency-related tools, the Green Deal is one of the vital financing mechanisms that enable householders and businesses to gain profits based on pay as-you-save policy. It encourages householders and business owners to install energy efficiency measures at no upfront cost and allow the payment to be made through their electricity bills. To become eligible for receiving finance aids, the savings must be greater or equal to the amount of the electricity bill. This policy came into force in January 2013 (Jonathan 2012). Another instrument is the Non-Fossil Fuel Obligation (NFFO) which mandates the former public electricity suppliers (PES) to purchase electricity from renewable generators. In fact, NFFO was introduced before the RO scheme introduction. It was the government's primary instrument of renewable energy policy. Currently, NFFO is no longer available to new generators; however, existing contracts are valid until the expiration of the last contract in 2019 (Ofgem 2016). Table 2.10 summarizes the most significant policies that have positively or negatively affected the introduction of renewable energy in the UK.

2.3.6 Renewable Energy Policies in Spain

Regardless of its limited hydrocarbons resources, Spain has successfully built a diverse and reliable power generation system after several years of effort. The primary energy consumed in Spain is a fossil fuel. Oil represents the dominant share of 42.3%, followed by natural gas at 19.8% and coal at 11.6%. Nuclear is accounted for up to 12%, while renewable energy sourcing 14.3% of the total energy consumed (Encyclopedia 2017). To maintain reliable and sustainable electricity supply, a great effort has been made to tackle the imbalance between the costs and revenues of the electricity generated from renewable resources. These reforms have a fundamental impact on improving the remuneration schemes facilitating access for a green economy. One step that significantly has an impact on renewable energy development in Spain is the

Table 2.10 A summary of actions taken by the government of the UK

Country	Total PV installed capacity (MW)	Effective actions taken by government	Ineffective actions taken by government
UK	13.108	Shifting from tendering to the quota system	Ambiguity of the number of projects the government intended to open for bidding has increased the risk for developers and investors
		Offering providers with secure long-time contract and finance	Absence of future guarantees increased the risk for investors
		Charging tax on all fossil fuels	Failing to provide a framework for stakeholders
		Dynamic pricing of Time-of-Use	The 5-year Non-Fossil Fuel Obligation contracts offered for winners were too long
		Obligating retailers to purchase renewable energy certificate which helps to create a new renewable market	Awarding projects to the lowest bidders negatively affected the industry

merge of the Civil Society Organizations (CSOs) and the independent developers as key advocate groups. The support of such groups was in line with the government's plan, gave a boost to renewable energy projects around the country (Karl 2012). As the renewable energy industry in Spain become more established in the 1999s, the roles these groups play have evolved to fill roles such as countering lobbies against renewable energy, leveraging required increase for renewable energy projects, and breaking policy or systemic deadlocks that emerge (Karl 2012).

In the mid-2000s, Spain introduced market premium and FiT incentive programs with very generous rates of remuneration schemes. This resulted in the most significant capacity ever installed in such a short time, making the Spanish solar power market in 2008 the largest market in the world. However, the government started to gradually reduce the uptake rates and financial incentives as a response to the economic challenges that the country countered. The dramatic increase in renewable energy installed capacity has posed a risk and distorted the electricity market. This urged Spain to suspend both market premiums and FiT at the beginning of 2012 and to become the first European country to take this action. It has been reported that the retroactive cuts to market premium and FiT incentives made by the government have posed a significant degree of risk that may negatively affect future investments in renewable energy projects (Karl 2012). RE incentives gave the project operators and developers who possess a capacity of 100 MW or less two options to choose from:

- To receive market premium incentives without an electricity purchase obligation
- To receive FiT incentives including a purchase obligation.

At the early stages of the FiT implementation, solar PV projects were guaranteed with a fixed rate for the first 25 years of the solar generating facility. However, for the reasons mentioned above, this fixed rate was reduced after that. In 2004, all solar PV projects which have a capacity less than 100 kW could receive a fixed tariff equivalent to 575% of the reference price, while projects with a capacity larger than 100 kW could receive a fixed tariff equal to 300% of the reference price. At that time, tariffs dedicated to PV generating facilities were receiving €0.40/kWh and €0.21/kWh, respectively. In 2008, FiT tariffs were modified and restructured to allocate additional incentives for rooftop solar PV systems (€0.34/kWh) and ground-mounted systems (€0.32/kWh) (Karl 2012). Regardless of all measurements that have been taken to readjust electricity rates, many users continued to pay rates that are, in fact, less than the cost of the all-in electricity, which resulted in an accumulating tariff deficit among Spanish utility companies.

Additionally, qualified renewable generating facilities that were receiving tariffs and market premiums that were higher than the market tariffs have also contributed to the tariff deficit (Brown 2013d). Even though financial subsidies are no longer available to new solar PV projects, it is a widely held view that solar PV systems will continue to be adopted in the future yet on a slower base (Brown 2013d). The Net Metering proposed in Spain in the form wherein energy generated by home-owners is purchased at a higher price than what is charged as consumers (Solaire 2007). Self-consumption (also known as Auto-generation) in Spain is similar to the Net Metering, yet in the former scheme, generation facility is connected inside the consumer's grid where the flow of energy from the consumer´s grid to the electrical system is not possible from a regulatory point of view. Under the self-consumption scheme, consumers are required to pay the same grid access tariffs and charges associated with the system costs as the rest of consumers (MedReg 2015). Table 2.11 presents the Spanish government's actions toward the implementation of renewable energy.

2.3.7 Renewable Energy Policies in India

In recent years, India put much effort into reshaping its economy and improving the average income rate of its citizens. It has already made notable progress at different levels. In conjunction with this development, India's energy sector has grown tremendously, along with economic and population growth (EIA 2016). India noticed a sense of urgency to develop new alternative energy sources and started to shift to a more diversified energy structure by offering a set of incentive programs. For instance, solar power has received plenty of financial incentives from the central and state governments of India, which resulted in a significant deployment of solar power across the

Table 2.11 A summary of actions taken by the government of Spain

Country	Total PV installed capacity (MW)	Effective actions taken by government	Ineffective actions taken by government
Spain	4.744	Capacity of less than 100 MW sell to distributors and distributors buy it at fixed rates	Weak policy changes caused an unwanted increase
		Guaranteed grid access for solar energy resources	Poor coordination between regional and national authorities
		Reducing the lifetime of FiT payment to 25 years	Supporting level is revised only every four years
		Providing generators with two options; a fixed feed-in rate adjusted annually or a fixed-premium rate on top of the electricity market	Complex administrative procedures and variation of regional procedures
			Allowing PV systems to connect to the low-voltage grid, while most generators have a medium voltage grid connection

country. Most of these efforts are dedicated to remote areas or areas that have electricity supply shortages. India has established the Indian Renewable Energy Development Agency acting under the Ministry of New and Renewable Energy (MNRE), where most of the initiatives dedicated to renewable energy technologies are initiated and managed (Behrendt 2015). As a means of catalyzing renewable projects, a 10-year tax deduction was offered to all project developers who are engaged in power generation and/or distribution activities assuming that they would begin before 31 March 2014. Based on its income, renewable generating facilities are obliged to pay a tax equal to 20–21% as a minimum alternative that would be offset thereafter (Behrendt 2015).

India has also focused on attracting foreign direct investment (FDI); the government has introduced the generation-based incentive (GBI) schemes as a means of supporting renewable projects under independent power producers (IPP). This initiative was designed for wind and solar power projects offering them accelerated depreciation. However, it was later found that FDI and IIPs did not apply to the depreciation provisions. The GBI tariffs for solar and wind energy projects which are fixed by the state regulatory commissions are available for all projects with a capacity up to 4000 MW and projects set up before March 2012 (Behrendt 2015). Several other measures such as RECs and renewable purchase obligation (RPO) and Net Metering were introduced by the National Action Plan on Climate Change (NAPCC) to promote renewable projects to meet the set targets of 10% by 2015 and 15% by 2020 at the national level.

RPO scheme obliges open-access consumers and distribution companies to purchase a certain share of their consumed energy from renewable energy resources.

Table 2.12 A summary of actions taken by the government of India

Country	Total PV installed capacity (MW)	Effective actions taken by government	Ineffective actions taken by government
India	26.869	Establishing independent ministry for nonconventional energy sources	Inadequate training and maintenance of the technology
		Introduction of renewable energy certificates (RECs)	Not developing adequate strategies to control the improper use of electricity by end-users
		Encouraging private investments in solar energy	Imbalance of tariff structures

The RECs' market started in 2011 to meet the RPO goals. However, procedures are being considered for modifying its provisions. Despite the criticism the RECs' mechanism receives for not being adequately supported yet, it is thought that the enforcement of RPO will help in securing the capacity needed for the RECs' market (Behrendt 2015). One of the most significant issues the solar sector in India encounter is inadequate training and maintenance of the field technicians, which caused the implementation of these applications to falter. Net Metering in India varies from state to state, but in general, consumers are allowed to sell the extra energy generated by their solar PV system to the grid and get compensated for the same (Bijli Bachao 2017). Unlike, for instance, the Net Metering proposed in Spain, wherein energy generated by home-owners is purchased at a higher price than what is charged as consumers (Solaire 2007). The government of India attempted to address the high initial cost of PV systems by offering 30% of the benchmarked installation upfront cost. In some states that lie in a special category, the subsidy can reach up to 70% of the benchmarked installation cost (Times Internet 2019).

Table 2.12 presents the most influential policies made by the government in India in an effort to catalyze the development of solar energy.

2.3.8 Renewable Energy Policies in Greece

Some significant reforms have been implemented in Greece aiming for energy efficiency improvement and further liberalization of the energy sector. It has been working with its neighboring countries to improve the integration of RES in the electricity market and to allegiance with prerequisites of EU state aid regulations. Greece has adopted ambitious goals and measures for developing the solar energy sector. The primary goal is to increase the contribution of RE in gross total final consumption by up to 20% by 2020. This number is triple the 7% share in 2005 and 2% higher than its EU obligation (Norton 2013). The system operators are obliged by law to offtake electricity generated at fixed prices. Priority dispatch of renewable energy is given to the system operator based on the grid code and other relevant standards offering 20-year power purchase agreements (PPAs). Renewable energy project developers, requesting to connect to the electricity grid, are obliged to agree on the PPA provision.

In Greece, there are two categories of FiT available for facilities generating power: FiT limited to solar PV installers and a FiT for non-solar PV renewable energy sources. Only the former is discussed in this research. FiT for solar PV installation came into force 2009 with separate tariffs for all other RE projects. Over time, fixed tariffs were gradually reduced. Tariffs for PV projects with an installed capacity of more than 100 kW located within the interconnected system, i.e., mainland Greece, were reduced to €292.08/MWh.

Moreover, tariffs for projects with an installed capacity of less than 100 kW were reduced to €328.60/MWh. On the other hand, other PV projects located out of the interconnected system are qualified to receive a tariff of €328.60/MWh with no limit of installed capacity. The government has also approved the VNM for specific investors, which enabled parties such as schools, universities, city, and regional councils, and farmers to develop solar PV in places other than the actual power consumption (Ilias 2013; Norton 2013). In 2013, the government set new retroactive measures to control the RES fund deficit issue caused by over-installment. The so-called new deal is included in a package of measures that oblige operating PV projects to accept feed-in-tariff cuts in exchange for extending the producers' bank loan duration and reducing their loan rates. It is thought that solar PV producers will be required to accept minimum FiT cuts of 25% for their operational assets. Moreover, it is also expected that specific details of each solar project, such as loan conditions and building cost, will also be taken into consideration and verified separately (Norton 2013). In Table 2.13, a summary of the most significant actions made by the government of Greece is presented.

Table 2.13 A summary of effective and ineffective actions taken by the government of Greece

Country	Total PV installed capacity (MW)	Effective actions taken by government	Ineffective actions taken by government
Greece	2.652	Encouraging technological and industrial research	Unprofessional response to opposing residents caused termination of some renewable energy projects
		Establishing a third-party financing scheme (pay not for the installation but the manufacture at a fixed monthly rate)	Long and complicated procedure directly affected the period of obtaining production and operation licenses
		Tax deduction of PV system cost	
		Modified building code	

2.4 Conclusion

This research aimed at summarizing seven selected countries' experiences in promoting solar power as an essential energy source in their energy mix. Whether it is political, technological, economic, or social, each one of these countries has faced several challenges and tackled them with different risks to create a new market for this new source of power. Most actions that positively or negatively affected the efforts of solar power deployment in each country are summarized.

As shown in Table 2.14, the shortcomings of each aspect were reviewed and classified based on four categories. The technological and financial aspects are critical, as they play an essential role in this equation and must not be overlooked. However, as can be seen in the table above *policy* and *management* aspects were crucial as well and significantly affected the implantation of renewable energy projects in most of the respective countries. Designing an optimal combination of all these areas will lead to a steady transition toward sustainable energy supply by utilizing this technology.

It is noteworthy that, unlike Saudi Arabia, all the solar energy leading countries are ruled by liberal democratic governments where governors are elected based on what they have promised to achieve for their electors. It is also important to note the clarity and the participatory-nature of the decision-making process in those countries which can help significantly in designing suitable policies by which the planned capacities are achieved. In democratic counties, such as the seven countries discussed above, public opinion is measured through elections, voting processes, and in some cases through demonstrations. Such practices do not exist entirely in Saudi Arabia. Thus, investigating the public opinion of energy options, and most importantly, RSPSs is imperative. These areas will be analyzed in both Chaps. 3 and 4.

As shown in Table 2.14, the unsuitable policy design and management and the lack of financial incentives are arguably behind the low adoption rates of RSPSs in Saudi Arabia. The structure of political systems is critical for any technology transition in any country. Saudi Arabia seeks this transition by developing new sectors and localizing new technologies. As a monarchy, the political system of Saudi Arabia differs from other similar patterns. For instance, the UK is a unitary parliamentary democracy and constitutional monarchy. The prime minister is the head of an elected parliament with wide-ranging authority to draft and pass regulations through the government. Saudi Arabia, however, tends to be more of a combination of 'Oligarchy' and 'Democracy' (Al-Turaiqi 2008), or "globalizing monarchy" (Henry and Springborg 2001). The King in Saudi Arabia is also the prime minister and his son is the Crown prince. All the ministers are appointed directly by the King. The Shura Council (Consultative Assembly), is the formal advisory of the country which has the authority to interpret and draft laws to the King, but not to pass them. All the members of the Shura Council are also directly appointed by the King. The lack of information about the government decision-making structure has always been

Table 2.14 Summative classification of ineffective actions taken by governments

Country	GDP/capita (US dollar)	Electricity consumption (TWh)	Total PV installed capacity (MW)	CO_2 Emission (Million ton)	Classification of ineffective actions Taken by Governments			
					Policy	Technology	Finance	Management
Saudi Arabia	20,760	274	35.2	562	●		●	●
Germany	40,997	534	45.930	753.63	●			●
Japan	32,486	935	55.500	1,207.78	●		●	
USA	55,888	3,868	25.620	5,485.74		●	●	●
UK	43,771	319	13.108	436.91	●			●
Spain	25,865	235	4.744	291.70	●			●
India	1.627	1050	26.869	1670	●	●		●
Greece	18,064	53	2.652	73.89	●			●

fueling speculations. To understand the opaque nature of the decision-making structure and process in terms of solar energy, an analysis will be carried out in Chap. 5 to add more clarity and capture potential factors that shape the solar energy policies in Saudi Arabia.

Further, it is important to note the policy design process in these countries is supported by a great deal of transparency and open information sources, which when compared to Saudi Arabia a great gap is recognized. As discussed in the Spanish case (Sect. 2.3.6), the CSOs have played a key role by positively influencing the development of renewable energy. It is unclear to what extent the CSOs and other non-business groups in Saudi Arabia are involved in the development of the solar energy industry. This area will also be discussed in Chap. 5.

The mechanism of Net Metering, for instance, has made great success in the US. On the other hand, a similar Net Metering mechanism that was introduced in Saudi Arabia turned out to be a big failure. As discussed above, the Net Metering model in the US is implemented throughout different flexible and supportive approaches. For example, consumers can select to send their excess generation to the next bill or receive a payment for credits. The Net Metering agreement in Saudi Arabia does not offer the latter option, nor does it offer to purchase surplus electricity at a rate higher than the wholesale rate. Subsidies for upfront costs are also not available for householders in Saudi Arabia, while the Government in India, for instance, pays 30% of the benchmarked installation cost and up to 70% of the benchmarked installation cost in some regions that lie in special categories.

How are the individuals, in a rentier monarchy such as Saudi Arabia, perceive energy options? what motivates them to or hinder them from accepting or rejecting to install RSPSs? and how are the solar energy policies designed in Saudi Arabia? To achieve the goal of this study of helping better understanding the current development of RSPSs, all of these areas will be shed light on and addressed through investigations carried out in the subsequent chapters.

2.5 Limitations

As stated above the goal of this chapter was to summarize the recent effective and ineffective policies taken by seven selected countries to build an understanding of the best practices of leading experiences. Thus, it does not involve a concrete level of analysis of these policies which might be considered as one of the limitations. Another limitation is excluding other countries which also considered leading countries in this field. A broader scope with more in-depth analysis will help to obtain more insights into this subject.

References

Al-Turaiqi, A. (2008). *The political system of Saudi Arabia*, 1st ed. ed. Ghainaa Publications.
Behrendt, L. (2015). Taxes and incentives for renewable energy. Congressional Research Service, The U.S.
BP. (2016). Statistical Review of World Energy, 1 St James's Square London SW1Y 4PD, UK.
Brown, P. (2013a). European Union Wind and Solar Electricity Policies: Overview and Considerations (No. 7–5700). Congressional Research Service, pp. 3–6.
Brown, P. (2013b). European Union Wind and Solar Electricity Policies: Overview and Considerations (No. 7–5700). Congressional Research Service, pp. 7–9.
Brown, P. (2013c). European Union Wind and Solar Electricity Policies: Overview and Considerations (No. 7–5700). Congressional Research Service, pp. 5–11.
Brown, P. (2013d). European Union Wind and Solar Electricity Policies: Overview and Considerations (No. 7–5700). Congressional Research Service, pp. 19–26.
California Public Utilities Commission, Net Energy Metering In California [WWW Document]. (2016a). Retrieved 16, Feb 2020, fromhttps://www.gosolarcalifornia.ca.gov/solar_basics/net_metering.php.
Crowe, S., Cresswell, K., Robertson, A., Huby, G., Avery, A., & Sheikh, A. (2011). The case study approach. *BMC Medical Research Methodology, 11,* 100. https://doi.org/10.1186/1471-2288-11-100.
Dokei, T., Arthur, M. M., Kazuo, K., & Yu, B. (2016). Changes to Japan's Existing Renewable Energy Feed-in-Tariff Systeml White & Case LLP [WWW Document]. Retrieved 12, Sep 2019, from https://www.whitecase.com/publications/alert/changes-japans-existing-renewable-energy-feed-tariff-system.
EIA [WWW Document]. (2016). Overview India, Analysis, U.S. Energy Information Administration (EIA). Retrieved 12, Sep 2019, from https://www.eia.gov/beta/international/analysis.php?iso=IND.
Electricity & Cogeneration Regulatory Authority, Small-Scale Solar PV Systems Regulations, 2017.
Encyclopedia [WWW Document]. (2017). Retrieved 12, Sep 2019, from https://www.encyclopedia.com/places/spain-portugal-italy-greece-and-balkans/spanish-and-portuguese-political-geography/spain.
EnergyCharts.de [WWW Document]. (2016). Retrieved 12, Sep 2019, from https://www.energycharts.de/energy.htm.
Energy.gov [WWW Document]. (2016). Leveraging Federal Renewable Energy Tax Credits. Retrieved 12, Sep 2019, from https://www.energy.gov/eere/slsc/downloads/leveraging-federal-renewable-energy-tax-credits.
Grimwood, G. G., & Elena, A. (2016). *Energy: The Renewables obligation*. UK.
Henry, C. M., & Springborg, R. (2001). *Globalization and the politics of development in the middle east*. The Contemporary Middle East: Cambridge University Press, Cambridge. https://doi.org/10.1017/CBO9780511807688.
Hertog, S. (2015). Oil prices: Eventually the Gulf states will run out of power [WWW Document]. Retrieved 30, Jan 2020, from https://theconversation.com/oil-prices-eventually-the-gulf-states-will-run-out-of-power-35867.
Ilias, T. (2013). Greece to pass a new fee for solar PV producers; legislate new metering—pv magazine International [WWW Document]. Retrieved 12, Sep 2019, from https://www.pv-magazine.com/2013/10/08/greece-to-pass-a-new-fee-for-solar-pv-producers-legislate-new-metering_100012974/.
Jonathan, R. (2012). The Green Deal [WWW Document]. Retrieved 12, Sep 2019, from https://www.mortgagefinancegazette.com/lending-news/the-green-deal-06-12-2012/.
Jordan, D. C., & Kurtz, S. R. (2013). Photovoltaic degradation rates—an analytical review. *Progress in Photovoltaics: Research and Applications, 21,* 12–29. https://doi.org/10.1002/pip.1182.
Karl, M. (2012). *Renewable energy policy and politics* (1st ed.). UK: Earthscan.

Krane, J. (2019). Energy Governance in Saudi Arabia: An assessment of the kingdom's resources, policies, and climate approach. In: *Energy PolicyMaking in a Cross-National Comparison: Energy Resources, Policy Processes and Law*. Baker Institute for Public Policy, Texa, U.S., p. 36.

Lindsey, H., Frontier Group, & Michelle, K. (2015). California's Solar Success Story How the Million Solar Roofs Initiative Transformed the State's Energy Landscape. Environment California Research & Policy Center.

Matthias, L. (2013). Ordinance Cutting Management Premium Pursuant to Renewable Energy Sources Act Promulgated. http://www.germanenergyblog.de/?p=11522.

MedReg. (2015). *Study to evaluate net metering systems in mediterranean countries*. Iyaly.

Mehran, K. (2013). *The modern middle east*. University of California Press.

National Conference of State Legislatures, State Net Metering Policies [WWW Document]. (2017). Retrieved 16, Feb 2020, from https://www.ncsl.org/research/energy/net-metering-policy-overview-and-state-legislative-updates.aspx.

Net Metering policy for roof top PVs in various states in India : Bijli Bachao [WWW Document]. (2017). Retrieved 16, Feb 2020, from https://www.bijlibachao.com/using-renewables/net-metering-policy-for-roof-top-pvs-in-various-states-in-india.html (accessed 2.16.20).

Norton, R. (2013). *European renewable energy incentive guide—Italy*.

NREL. (2015). Renewable Portfolio Standards Resources & Technical Assistance. the U.S.

Ofgem [WWW Document]. (2016). Non Fossil Fuel Obligation (NFFO)/ Scottish Renewable Obligation (SRO). Retrieved 12, Sep 2019, from https://www.ofgem.gov.uk/environmental-programmes/nffo.

Oxford Dictionaries [WWW Document]. (2017a). Lexico Dictionaries, technology. Retrieved 9, Dec 2019, from https://www.lexico.com/en/definition/technology.

Oxford Dictionaries [WWW Document]. (2017b). Lexico Dictionaries, policy. Retrieved 9, Dec 2019, from https://www.lexico.com/en/definition/policy.

Oxford Dictionaries [WWW Document]. (2017c). Lexico Dictionaries, finance. Retrieved 9, Dec 2019, from https://www.lexico.com/en/definition/finance.

Oxford Dictionaries [WWW Document]. (2017d). Lexico Dictionaries, management. Retrieved 9, Dec 2019, from https://www.lexico.com/en/definition/management.

Planète, E. (2015). The History of Energy in the United Kingdom [WWW Document]. https://www.planete-energies.com/en/medias/sagaenergies/history-energy-united-kingdom (accessed 12.9.19).

Photovoltaïque connexion réseau EDf : Vente et installation de panneau solaire photovoltaique [WWW Document]. (2007). Retrieved 16, Feb 2020, from https://web.archive.org/web/20070208220615/http://www.bati-depot.fr/solaire/photovoltaique/electricite-solaire/photovoltaique-connexion-reseau.html.

Renewable Energy Project Development Office. Ministry of Energy, Industry and Mineral Resources [WWW Document]. (2017). Retrieved 21, Aug 2019, from https://www.powersaudiarabia.com.sa/web/index.html.

Saudi Electricity Company, consumption tariffs [WWW Document]. (2018). Retrieved 30, March 2020, fromhttps://www.se.com.sa/en-us/customers/Pages/TariffRates.aspx.

Saudi Press Agency [WWW Document]. (2019). Retrieved 23, Nov 2019, from https://www.spa.gov.sa/viewstory.php?lang=en&newsid=1963700.

Saudi Vision [WWW Document]. (2016b). Retrieved 1, Jan 2020, from https://vision2030.gov.sa/en/node/87.

SEIA Solar Energy Industries Association, Solar Industry Research Data, 2015.

TNS. (2019). True News Source. Retrieved 23, Nov 2019, from https://truenewssource.com/2019/08/31/king-salman-royal-decree-gives-minor-jolt-to-khalidal-falih/.

Turki Alshehri, 2019. (2019). Future projects for renewable energy in the Kingdom, a presentation given in Riyadh Chamber of Commerce, Riyadh, Saudi Arabia.

Zainal, Z. (2007). Case study as a research method. *Jurnal Kemanusiaan*, 5.

Chapter 3
An Exploratory Study of the Public's Views on Residential Solar Photovoltaic Systems in Hydrocarbon-Rich Saudi Arabia

In this chapter, an empirical exploratory investigation of the individual households' perception of the environmental situation and energy option was conducted. Using a non-probability convenience sampling, a cross-sectional web-based survey was employed covering the five main areas in Saudi Arabia (Central, Northern, Eastern, Southern, and Western areas). The questionnaire is divided into two parts; the first part is presented in this chapter, while the second part of the questionnaire is presented in Chap. 4. The first part of the questionnaire, which presented in this chapter, mainly focuses on five aspects, including basic demographic characteristics, residential status, energy consumption, environmental and energy cognition, and willingness to adopt RSPSs.

3.1 Research Background

With the increasing threats of climate change and the depletion of fossil fuels, the use of alternative renewable energy sources has become a priority in many countries. In the Middle East, solar energy represents a huge investment opportunity, and several Arab countries have announced ambitious solar energy targets over the past decade (REN21 2017). Reliance on liquid fuels to meet domestic demand has pushed countries like Morocco, Jordan, the UAE, and Saudi Arabia to diversify their energy sources using various approaches (REN21 2017). This is especially true for the oil-rich Arab states of the GCC. Saudi Arabia, the largest country among the GCC states, is aiming to reshape its national income resources by developing alternative sectors to secure a more sustainable and diversified economy (Saudi Vision 2016a, b). In a country where abundant fossil fuel is the main source of energy, subsidized energy has led to excessive use of energy in Saudi Arabia, exhausting a substantial portion of

The original version of this chapter was revised: figure 3.7 has been corrected. The correction to this chapter is available at https://doi.org/10.1007/978-981-15-5803-0_7

K. A. Alrashoud, *Managing Technology Transition in Saudi Arabia*,
https://doi.org/10.1007/978-981-15-5803-0_3

the fossil fuel reserves that could have been exported (Alshehry and Belloumi 2015; Woertz 2013). In conjunction with rapid population growth, electricity consumption has reached unprecedented levels, especially in the residential sector, which is responsible for over 50% of total electricity consumption (Tlili 2015). To ensure sustainable development and maintain its position as one of the world's largest energy suppliers, it is vital for Saudi Arabia to actively use renewable energy technologies, including RSPSs.

RSPSs are small-scale applications that can be installed on a rooftop or in any available residential area to be exposed to sunlight and generate electricity. This type of system usually consists of solar panels, an inverter, a switchboard, an electricity meter, and a storage battery through which electricity can be provided at night when needed. Among all forms of renewable technologies, solar PV has recorded the highest growth rates, constituting over 33% of the annual capacity in 2016 (REN21 2017). Although the utility-scale projects represent a significant share of the solar market in many parts of the world, small-scale application projects are increasing rapidly.

In March 2018, Saudi Arabia and the SoftBank Group announced their collaboration to develop a 150–200 GW solar project (Arthur D Little 2018). However, the capacity of small-scale PV systems seems to be excluded from this massive project (Al Kobraish 2018; Al-Taher 2018). Even though this massive capacity is expected to be added to the current 87.7 GW installed power capacity (Arthur D Little 2018), the involvement of the residential sector is still crucial for many reasons. The Saudi government has approved a gradual elimination of the subsidies applied to energy prices. In January 2018, the electricity tariff was almost tripled, and many other subsidies were eliminated from other energy forms.

Given that the electricity tariffs will continue to increase until global standards are met (Almusabihi 2018), this represents a challenge for the residents of Saudi Arabia. Especially when the average Saudi GDP per capita is USD 20,796 compared, for instance, with the United States of America's GDP of $53,128 US dollars (Trading Economics 2018). Further, a portion of the power generated from the enormous 150–200 GW solar project will be exported to neighboring countries (SEC 2018). Moreover, Saudi Arabia's economy is projected to experience gradual growth, and many substantial projects that demand a significant amount of power are expected to be completed in the coming years. Examples of these impending projects include the Riyadh metro, with 176 km of electric system length, the Neom smart city, and the Red Sea megaproject (ADA 2018; T-Systems 2018). Saudi Arabia recognizes the importance of involving individual households in electricity generation. Thus, it has designed primary regulations to manage the transmission of electricity between the related parties (Al-Barrak 2018).

In August 2017, ECRA issued a regulatory framework for the end-users to operate their own RSPSs and export excess power to the national grid (ECRA 2017). The regulatory framework was intended to come into play on July 1, 2018; however, no further details or incentive programs have been announced since the writing of this book. SEC has called upon individuals wanting to integrate RSPSs at their residence units to submit their application, but the number of applications received is reportedly small, falling short of expectations (Al-Taher 2018). This ambiguity is compounded by a complete lack of official studies regarding the public's views on the installation of RSPSs in Saudi Arabia (Al-Barrak 2018; Al-Taher 2018).

Public perception is influenced and shaped by many different factors. Having a robust understanding of what motivates or hinders public opinion is key to formulate solid policies and regulations to ensure the successful adoption and integration of RSPSs into the power grid. Considering both the importance of each individual household's role in the energy scenario of Saudi Arabia and the critical role of social perception in the adoption of new technologies, this research aims to explore the Saudi Arabian public's views on the adoption of RSPSs. The paucity of large-scale studies investigating and analyzing this issue, in this context, emphasizes the significance of this study.

3.2 Social Acceptance of RSPSs

In recent years, attention has been devoted to the social perceptions surrounding the acceptance of renewable-based energy as alternative sources of power. Social acceptance is defined as a positive attitude toward a particular technology or measure, and this drives the acceptance of a supporting action, if needed (Edison and Geissler 2003; Huijts et al. 2012a, b). The author of this book realizes that some articles distinguish between "acceptance" and "acceptability," where the former is indicative of supportive behavior toward object or situation, and the latter indicates an attitude toward an object or situation (Huijts et al. 2012a, b). In this research, however, the term "acceptance" is defined as: "agreement with or belief in an idea or explanation (Oxford Dictionaries 2012) by exhibiting a favorable attitude or positive perception."

Human choice is a critical and controlling factor in energy use, and it is influenced by various contextual and personal circumstances (Abdul Hai 2019). Similar to the other GCC member states, the culture in Saudi Arabia is formed by central elements such as religion, tribal system, and the nature of the political regime. Regarding religion, Saudi Arabia is occasionally referred to as the "home of Islam" and has occupied a special place among Muslim countries because it is home to the Two Holy Mosques of Islam (Chaplin 2014). Thus, religion fundamentally influences the social norms, life patterns, attitudes, and everyday practices in this society (AL-Shehry 2006). The religion of Islam calls for moderation and sobriety, and it emphasizes the significance of the conservation of natural resources, as well as environmental protection (Ayaz et al. 2003; Erhun 2001).

Further, the existence of a tribal system is still evident in Saudi society today, and this system impacts the status of individuals within society and could affect their choices (Aldraehim et al. 2013; AL-Shehry 2006). The adoption of new technologies in Saudi Arabia is often impacted by the importance of family and traditions, which are emphasized by the local culture (Aldraehim et al. 2013). Social norms, along with social awareness, economic viability, and political responsibility, must coexist to achieve the goal of successful integration of renewable-based energy (IRENA 2013).

The growth of the installed RSPSs capacity has assisted many countries in achieving their goals of developing a diversified energy portfolio. However, this development is not achieved without challenges. Based on a survey conducted by Claudy et al. (2010), several potential barriers to the adoption of RSPSs exist in the Republic of Ireland. These barriers include the upfront investment required to purchase an RSPS, the interference of the RSPS in day-to-day lifestyle, the difficulty of understanding and using the RSPS, and the degree to which the RSPS can be tested prior adoption were the most significant factors hindering Irish homeowners from installing solar PV systems. Similar to any other long-term investment, uncertainty is a critical factor challenging the acceptance of RSPSs adoption.

A study by Sütterlin and Siegrist (2017) investigated the public's acceptance of different renewable energy technologies in Switzerland by interviewing a total of 1211 participants. Regarding the solar power analysis in their study, Sütterlin and Siegrist used the continued association technique to assess participants' affective imagery of solar power at both of general and concrete level. They found that "this effective image analysis using word association allows the identification of positive and negative connotations and symbolic meanings associated with a stimulus in an unfiltered, relatively context-free, and spontaneous way." In their study, interviewees were asked general questions about solar energy, and then they wrote down the first word or thoughts that came to their minds. Afterward, the affective value of the association was evaluated by asking the interviewees to answer what best describes their thoughts on an 11-point scale ranging from −5 for "extremely negative" to +5 for "extremely positive." The results of the analysis indicated that individuals do not think about drawbacks related to renewable energy when they consider it from a general perspective. The results also showed that the public acceptance of renewable energy technologies decreases when presented with the drawbacks or obstacles associated with these technologies (Sütterlin and Siegrist 2017).

In another study conducted by Siegrist's group (Visschers and Siegrist 2014), in Switzerland, public acceptance of five energy technologies (including solar power, wind power, nuclear power, hydro-power, and gas-fired power), along with potential determinants factors were investigated. Data were collected through randomly sending mail questionnaires to addresses in the German-speaking part of Switzerland. The adult in the household whose birthday was soonest was invited to complete the survey. This study revealed that acceptance was highest for solar power and hydro-power, and the protected values, perceived cost, and perceived benefits were significant determinants in explaining the public's acceptance of energy technologies in Switzerland. Unexpectedly, Visschers and Siegrist observed that gas-fired power was rated more negatively than nuclear power. They attribute this results to three potential factors: Swiss people's low familiarity with gas-fired power, CO2 emission of this type of energy resource, and use of gas-fired power plants would make Switzerland more depended on other foreigner countries. However, the mail and telephone approaches may not have represented the population accurately. The study found that more people who had completed higher secondary school and fewer people who had only completed compulsory education were included in the sample when it was compared to the Swiss population as a whole. The sample also included a higher proportion of men and older people in comparison to the actual demographics of Switzerland. The cause is believed to be the selected approach since older people are more likely to be listed in the telephone directory than younger people (Visschers and Siegrist 2014).

The involvement of the residential sector in generating electricity through the integration of RSPSs is critical for Saudi Arabia because the residential sector accounts for over half of the total energy consumption (Tlili 2015). While a considerable number of studies related to the implementation of solar energy in Saudi Arabia have been conducted, only a limited number of them have discussed the perception of end-users toward installing RSPSs in households (Sect. 1.2). This is not abnormal, given the fact that the pact for the first independent solar power project with a capacity of 50 GW was only signed in 2016, and the regulations of small-scale solar PV systems did not even come into play until July 2018 (ECRA 2017; Renewablesnow 2016).

3.3 Research Questions and Hypotheses

With the aim to explore the public's views in Saudi Arabia on the adoption of RSPSs, and in keeping with the focus on the facets being examined in this study, the following research questions and hypotheses were developed and tested by using empirical data.

3.3.1 Research Question

RQ1. How individuals in Saudi Arabia perceive energy options?

3.3.2 Research Hypotheses

(H1) There is a specific preference among the Saudi population against using crude oil.

(H2) The majority of the Saudi population is willing to install RSPS if they would receive subsidies equivalent to 40% of the total cost.

3.4 Methodology and Research Design

The present study was conducted through a cross-sectional web-based survey covering the five main areas in Saudi Arabia; these include the Central, Northern, Southern, Eastern, and Western areas. Because the questionnaire was designed to serve as an exploratory study of the residents' views in Saudi Arabia toward RSPSs, a cross-sectional analysis was used because of its advantages in evaluating prevalence, more than longitudinal analysis or other types of observational research methods. Further, frequencies and proportions were calculated to describe the extent of public acceptance of RSPSs in Saudi Arabia.

3.4.1 Sampling Method

The present study used a non-probability convenience sampling approach attempting to represent the decision-making capacity of the Saudi community. Despite the limitations of a web-based approach, such as the need for access to the internet or registered email, a web-based approach was selected in this case because it appeals to a much wider audience under the same time and cost constraints (Ilieva et al. 2002). A web-based approach has several advantages over face-to-face questionnaires because, generally, respondents are willing to take the time to complete a questionnaire, while the face-to-face approach may pressure them to do so. This is especially true if the questionnaire requires a long time to complete (Bryman 2012).

The survey was distributed through several online channels, including e-mails, LinkedIn, WhatsApp, Facebook, and Twitter, during the period of March-May 2018. Through this questionnaire, 1500 responses were collected with two missing interactions, resulting in 1498 collected responses, allowing a <1% error rate at a 95% confidence level.

Considering the large proportion of youth in the Saudi population (46.3% are < 40 years old), which also ranks first worldwide in smartphone use, a web-based survey was selected as the most appropriate approach to conduct this study (Clayton 2012; GAS 2018; MobileGate 2017). The author did not set any restrictions on nationality, age, gender, or socioeconomic class. Consequently, the contributions from the younger age groups were more compared to other groups because they tended to be connected online more than the other age groups. The author also did not limit the number of responses to a single device because it was recognized that some respondents could be sharing the same device. The author exerted many efforts to avoid selection bias, making the survey available in all trending social media groups, including region-specific groups. In contrast, the author avoided channels such as environmental protection and academic groups related to renewable energy because they were expected to be biased in favor of solar energy use.

The author used a chi-squared test, the Fisher exact test (for expected values of less than 5), and corresponding *p*-values were generated during analysis. Statistical significance was determined for *p*-values less than 0.05, with a consequent analysis of the strength of the association performed using odds ratios and regression modeling. The reliability testing was ranked as 'acceptable' to 'good' (DeVellis 2016), where Cronbach's alpha ranged from 0.735 to 0.814 for the different scales included in the study. The questionnaire was reviewed by two experts in the field, and its validity was tested by selecting a pilot sample, which contributed to the design of the questionnaire.

3.4.2 *Questionnaire Design*

To achieve the goals of this study, the questionnaire was divided into five sections, including basic demographic characteristics, residence status, household energy use, environmental and energy cognition, and willingness to adopt RSPSs.

In the basic demographic characteristics section, respondents were asked to answer closed-ended questions by choosing the answers that best match their demographic situation, including nationality, gender, location of the residence, age, occupation, education, and income (Table 3.1).

The residence status section described the ownership, type, and age of the residence, as well as the likelihood to move to another residence in the future. All the questions of this section were closed-ended questions where respondents had to choose one answer for each question (Tables 3.2 and 3.3).

The household energy use section assessed the participant's perception of the recent electricity tariff hike and their electricity consumption behavior. This was done using the 5-point Likert scale, such that 1: Very low; 2: Low; 3: Moderate; 4: High; and 5: Very high, where lower values indicate weaker agreement with the statement and higher values indicate stronger agreement with the statement. All questions assessing the extent of agreement/disagreement or satisfaction for a particular statement were assessed on this same 5-point Likert scale (Table 3.4).

Questions regarding the environmental and energy cognition section discussed the participants' perception of global warming, energy supply, renewable energy, and their willingness to contribute to tackling the challenges associated with these topics.

In the last part of the questionnaire, all the respondents were asked to answer two further dichotomous questions, to gain insight into their willingness to adopt RSPSs. Based on their answer to the suggested proportion of subsidy in the first question, both acceptors and rejectors were directed to answer the same questions. However, the questions had a different hypothetical proportion of subsidies based on a respondent's previous answers, where the acceptors were offered a lesser subsidy (20%), and the rejectors were offered a larger subsidy (60%).

3.5 Results

3.5.1 *Basic Demographic Characteristics*

Analysis of the respondents' demographic characteristics indicated that the majority of respondents were Saudi nationals (94.8%), compared to only (5.2%) of non-Saudi respondents. In terms of respondents' gender, male participation was dominant at 73.8%, compared to 26.2% of female respondents. Respondents engaged in the questionnaire from all the main five areas covered in this study, and the Central region showed the largest participation (49.7%), followed by the Western (23.6%), Eastern

Table 3.1 Frequencies and proportions of RSPS acceptance by demographic characteristics ($n = 1498$)

Characteristic		Frequency	Percent (%)	Proportion accepting RSPS (%)	p-value (For differences in RSPS acceptance)
Nationality	Saudi	1420	94.8	97.5	1.000[b]
	Non-Saudi	78	5.2	97.4	
Gender	Males	1139	73.8	97.1	0.114
	Females	359	26.2	98.6	
Residence location	Central area	744	49.7	97.6	0.105[b]
	Eastern area	232	15.5	97.0	
	Southern area	96	6.4	96.9	
	Western area	353	23.6	98.6	
	Northern area	73	4.9	93.2	
Age	18–30	495	33.0	98.8	0.101[b]
	31–40	531	35.4	96.8	
	41–50	259	17.3	96.1	
	51–60	167	11.1	98.2	
	61–70	43	2.9	95.3	
	71 or more	3	0.2	100	
Occupation	Student	244	16.3	99.6	**0.015**[a, b]
	Government employee	524	35.0	97.1	
	Private sector employee	339	22.6	96.8	
	Self-employed	131	8.7	94.7	
	Retired	152	10.1	97.4	
	Other	108	7.2	100.0	
Education	Postgraduate	274	18.3	98.9	0.067[b]
	Undergraduate	874	58.3	96.7	
	Secondary or equivalent	314	21.0	98.7	
	Junior high school	29	1.9	93.1	
	Elementary school	6	0.4	100.0	
	Informal education	1	0.1	0	
	Not gone to school at all	1	0.1	100.0	

(continued)

Table 3.1 (continued)

Characteristic		Frequency	Percent (%)	Proportion accepting RSPS (%)	*p*-value (For differences in RSPS acceptance)
Income (SAR)	Less than 5000	117	7.8	96.6	0.747[b]
	5000–9000	302	20.2	97.4	
	10000–14000	350	23.4	96.3	
	15000–19000	242	16.2	97.9	
	20000–24000	192	12.8	97.4	
	25000–29000	85	5.7	98.8	
	30000–34000	63	4.2	100.0	
	35000–39000	41	2.7	97.6	
	40000 or more	106	7.1	99.1	
All participants		**1498**	**100**	**97.46**	

[a]Statistically significant results (p-value < 0.05)
[b]For these characteristics, a Fisher exact test was used because some of the expected values were lower than 5, making the chi-squared test results less accurate

(15.5%), Southern regions (6.4%). The Northern showed the lowest participation (4.9%). Respondents in this study spanned different age groups and saw a higher proportion of young adult respondents (< 40-year-old). As shown in (Table 3.1), the age group between 31 and 40-year-olds was the largest share of the sample size representing 35.4%, and 18–30-year-olds followed this group at 33%. The age group of 41–50-year-olds represented 17.3%, followed by the 51 to 60-year-olds, 61 to 70-year-olds, and >71-year-olds at (11.1%), (2.9%), and (0.2%), respectively. The demographic characteristics of the respondents indicated that the majority of respondents were young adult males (<40-year-old) who were employed (governmental or private sectors) with undergraduate education and who had monthly incomes of less than SAR 14,000 (USD 3,733) as shown in (Table 3.1). The results of the dichotomous question, "Do you accept or reject the notion of installing an RSPS in your dwelling?",[1] indicating that the dominant percentage of respondents (97.46%) was in acceptance of RSPSs (95% CI[2]: 96.7% to 98.3%). Moreover, analysis of all targeted areas (Central, Eastern, Southern, Western, and Northern areas) showed no significant variation in acceptance or rejection rates by region (Table 3.1).

[1]This is a common question between in this chapter and Chap. 4 to which most of the results of other questions are compared.

[2]The abbreviation of 'CI' herein stands for Confidence Interval.

Table 3.2 Numbers and proportions of acceptance of RSPSs by residence status ($n = 1498$)

Characteristic		Frequency	Proportion (%)	Accept RSPS (%)	*p*-value (with RSPS acceptance)
Ownership the current residence	Yes	783	52.3	96.9	0.173
	No	715	47.7	98.0	
Residence status	Self-built	580	38.7	96.2	0.025[a, b]
	Purchased	265	17.7	98.9	
	Mortgage	51	3.4	96.1	
	Renting	538	35.9	98.5	
	Other	64	4.3	95.3	
Residence type	Apartment (Flat)	555	37.0	97.7	0.645[b]
	Villa	724	48.3	97.4	
	Semi-detached	104	6.9	96.2	
	Apartment complex	25	1.7	100.0	
	Cottage	74	4.9	98.6	
	Other	16	1.1	93.8	
Residence age (Year)	1–5	423	28.2	96.5	0.182[b]
	6–10	337	22.5	97.0	
	11–15	192	12.8	96.4	
	16–20	187	12.5	98.9	
	21 or more	302	20.2	98.7	
	Don't know	57	3.8	100.0	
Likelihood of translocating in the next ten years[c]	Not at all likely	102	9.1	95.1	0.058[b]
	Slightly likely	217	19.4	98.2	
	Neutral	413	36.9	96.4	
	Quite likely	247	22.1	97.6	
	Extremely likely	141	12.6	100.0	

[a]Statistically significant results (p-value < 0.05)
[b]For these characteristics, a Fisher exact test was used because some of the expected values were lower than 5, making the chi-squared test results less accurate
[c]The total number of respondents answering the question concerning the likelihood of translocating was 1120 because the data for 370 respondents were missing

Table 3.3 Association between renting and the likelihood of translocating in the next ten years

Likelihood of translocating in the next ten years	Number	Proportion renting (%)	*p*-value
Not at all likely	102	8.8	<0.001[a, b]
Slightly likely	217	14.3	
Neutral	413	33.9	
Quite likely	247	48.6	
Extremely likely	141	67.4	

[a]Statistically significant results (p-value < 0.05)
[b]Highly significant (p-value < 0.01)

3.5.2 *Residence Status*

Nearly half of the respondents owned their homes. Also, the proportion of participants who had built their homes were comparable to those who were renting, where the proportion was close to one third for each category. Moreover, half of the respondents were living in villas, and the other half of the respondents' residences were less than ten years old. The likelihood of translocation displayed a bell-shaped distribution that is slightly skewed toward participants who were more likely to translocate within the next ten years. The bivariate analysis of residence profiles in acceptance of RSPSs indicated no statistically significant association with all variables except for the residence situation. Self-built and mortgage respondents had higher rejection rates compared to purchasing and renting respondents.

Conversely, a lower RSPS rejection rate was observed for those who were more likely to translocate in the next ten years (Table 3.2). Although the results slightly fail from the significant cut off point ($p = 0.058$), approximately 25% of respondents did not answer this question, which may have contributed to the non-significant results. The residence situation was also significantly associated with the likelihood to translocate, where the proportion of participants renting their homes increased significantly with the increase in the likelihood of translocating in the next ten years (Tables 3.2 and 3.3). However, there was no statistical significance for both variables when they were included in the multivariate regression analysis (Table 3.6).

3.5.3 *Household Energy Use*

The data indicated a general trend that disagreed with the recent energy price hike (Table 3.4). About half of the respondents had monthly electricity bills that ranged between SAR 200 and 700 (USD 50 and 187). Also, half of the respondents did

Table 3.4 Frequencies and proportions of acceptance of RSPSs by household energy use (n = 1498)

Characteristic		Frequency	Percent (%)	Percent accepting RSPS (%)	p-value (with RSPS acceptance)
Satisfaction level with recent energy prices raise	Strongly agree	50	3.3	96.0	0.433[c]
	Agree	134	8.9	98.5	
	Neutral	305	20.4	97.7	
	Disagree	403	26.9	98.3	
	Strongly disagree	606	40.5	96.7	
Electricity monthly bill (SAR)	200 or less	208	13.9	98.6	0.413[c]
	200–700	724	48.3	97.9	
	700–1200	318	21.2	95.9	
	1200–1700	119	7.9	96.6	
	1700–2300	56	3.7	94.6	
	2300–2800	27	1.8	100.0	
	2800–3300	23	1.5	100.0	
	3300–3800	7	0.5	100.0	
	3800 or more	16	1.1	100.0	
Frequency of checking electricity consumption[d]	Weekly/Biweekly	24	2.4	87.5	**0.006**[a, b, c]
	Monthly	247	27.0	96.8	
	Quarterly	91	10.2	98.9	
	Twice a year	30	3.2	93.3	
	Annually	7	0.8	100.0	
	Very infrequently	97	10.5	95.9	
	Not at all	411	46.0	99.0	
Recent energy saving measures	Yes	1030	68.8	97.9	0.143
	No	468	31.2	96.6	

[a]Statistically significant results (p-value < 0.05)
[b]Highly significant (p-value < 0.01)
[c]For these tests, the Fisher exact tests were used because some of the expected values were lower than 5, making the chi-squared test results less accurate
[d]The total number of respondents for the frequency of checking electricity consumption was 885, while the data of 613 respondents were missing. As these missing data have no relation to other survey variables, their occurrence is considered "Completely at Random" and does not affect the results of the other survey questions

not check their electricity consumption, while a quarter of them checked their electricity consumption monthly. In contrast, the majority of respondents had recently implemented energy-savings measures.

The bivariate analysis for the energy use variable with RSPS acceptance indicated no significant association among all variables except for the frequency of checking electricity bills, which was highly significant ($p = 0.006$, Table 3.4). The highest RSPS rejection rate was for those who checked their electricity consumption once or twice weekly (12.5%). This rate is also very high compared to the rejection rates of all bill price categories.

The analysis of the correlation between income and frequency of checking electricity consumption indicated that there is a statistically significant difference in the distribution of the participants by these two variables ($p = 0.002$). Each indicator reflects different information related to the participants' economic situation (Table 3.5).

3.5.4 Environmental and Energy Cognition

The vast majority of respondents, 87.2%, had previously heard about the concept of global warming, while 6.8% answered that they might have heard about global warming, and 6% said they had not heard about global warming previously. Additionally, most of the respondents (75%), strongly agreed or agreed that the main cause of global warming was the result of fossil fuel emissions from coal, petrol, and diesel. At the same time, a neutral trend was observed on whether these effects were considered to be exaggerated or not (Fig. 3.1).

When the participants were asked the question, "What do you think is an appropriate approach to address global warming as it is related to energy production?", the respondents were given seven options from which they had to select only one option. The respondents had a high level of awareness regarding issues related to climate change and a desire to expand the utilization of renewable energy. Among the answers, 65% favored expanding renewable power usage to address global warming impacts, while 6%, 7%, and 11% believed that the "expansion of nuclear power," "reduction of electricity consumption," and "investment in research and development" was the appropriate approach, respectively. Merely 3% of the participants thought that there is no need to take action to tackle climate change, and just 2% thought that there is no threat perceived by global warming. The remaining 6% answered that they do not know which approach is appropriate to mitigate climate change (Fig. 3.2).

While the majority of respondents recommended the expansion of renewable energy to address the effects of global warming, when respondents were asked, "If it solved global warming, would you be willing to pay ___ more per month on your electricity bill (in SAR)?", the majority (65%) were not willing to pay more than SAR 10 (USD 2.5) monthly to fight global warming. A proportion of 10.1% of respondents showed their willingness to pay SAR 11–29 extra (USD 2.9–7.7) on their electricity bills. Similarly, 9.1%, 7%, and 8.6% of respondents were willing to pay SAR 30–49 (USD 8–13), SAR 50–69 (USD13.3–18.4), and SAR 70 (USD 18.6) or greater, respectively (Fig. 3.3).

Table 3.5 Correlation between monthly income and frequency of checking electricity consumption

p-value = 0.002**		Frequency of checking electricity consumption[2]													
		Weekly/Biweekly		Monthly		Quarterly		Twice a year		Annually		Very infrequently		Not at all	
		Count	Percent (%)	Count	Percent (%)	Count	Percent (%)	Count	Percent (%)	Count	Percent (%)	Count	Percent (%)	Count	Percent (%)
Income (SAR)	Less than 5000	1	1	34	42	10	12	1	1	0	0	6	7	29	36
	5000–9000	7	4	49	27	14	8	4	2	1	1	17	9	91	50
	10000–14000	9	4	58	28	24	12	4	2	4	2	21	10	88	42
	15000–19000	5	3	43	28	13	8	4	3	0	0	22	14	66	43
	20000–24000	1	1	29	25	13	11	6	5	0	0	13	11	52	46
	25000–29000	1	2	13	25	8	15	1	2	0	0	6	12	23	44
	30000–34000	0	0	6	19	1	3	3	9	0	0	0	0	22	69
	35000–39000	0	0	2	7	1	4	2	7	2	7	6	21	15	54
	40000 or more	0	0	13	23	7	13	5	9	0	0	6	11	25	45

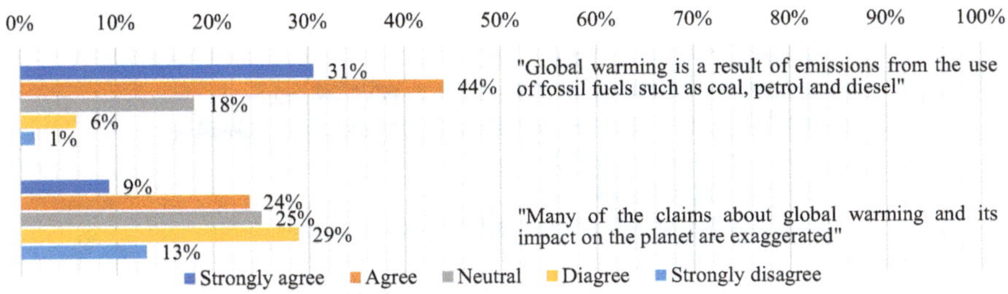

Fig. 3.1 Respondents' knowledge of global warming causes and effects

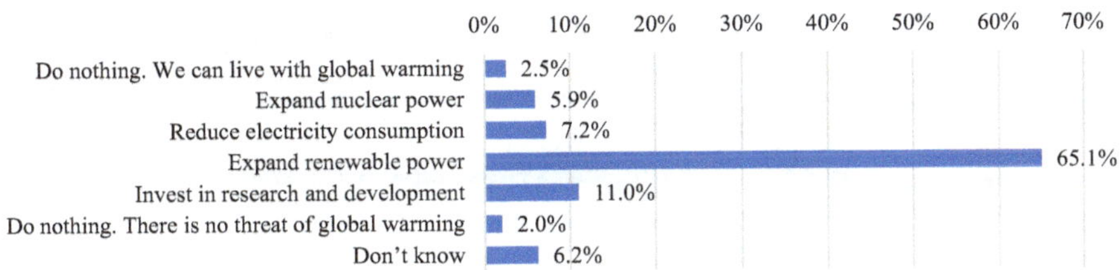

Fig. 3.2 Respondents' recommendation on global warming

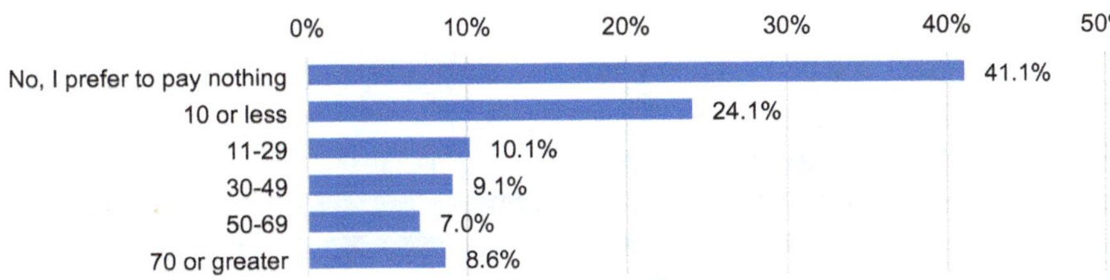

Fig. 3.3 Respondents' decisions on paying extra money for electricity monthly bill to address the effects of global warming

A dissimilarity was observed among the respondents' expectations regarding the longevity of crude oil in Saudi Arabia. While 10.3% and 19.3% of the respondents expected the crude oil to last for 10 or 25 years, respectively, the remaining 70.4% expected that crude oil would last between 50 to more than 100 years (Fig. 3.4).

As shown in (Fig. 3.5), the results show a general preferential trend for changing from oil dependence to other energy alternatives. Most respondents think that there are available alternatives to oil and do not prefer oil to continue as the primary energy source. They also acknowledge the existing problems of air pollution and climate change. Inference from these results confirms that the majority of the Saudi population (71.9%) are not in favor of crude oil to continue as the main energy source for the country (95% CI: 69.6% to 74.2%).

Regarding the question of when the respondents first heard about solar energy, the graph shows this general trend of increasing knowledge about solar energy to

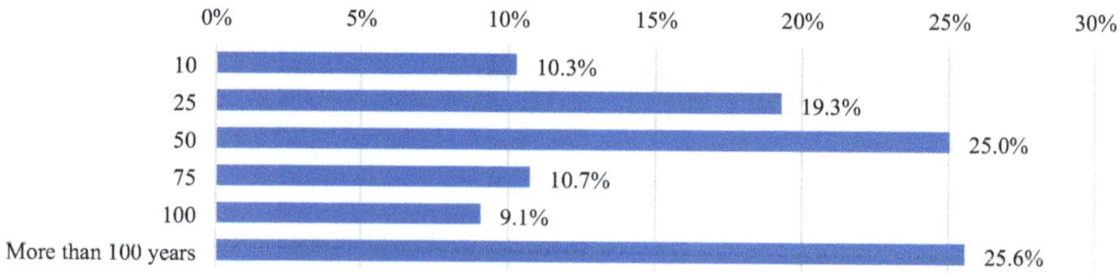

Fig. 3.4 Respondents' expectations on the longevity of crude oil in Saudi Arabia

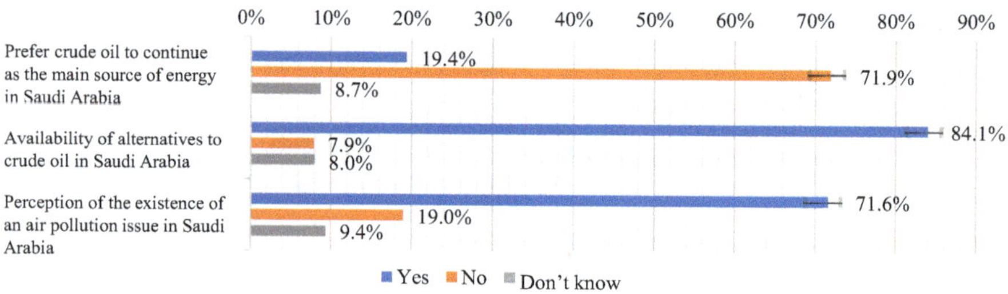

Fig. 3.5 Respondents' preferences on crude oil as a source of energy and perception of the existence of an air pollution issue in Saudi Arabia

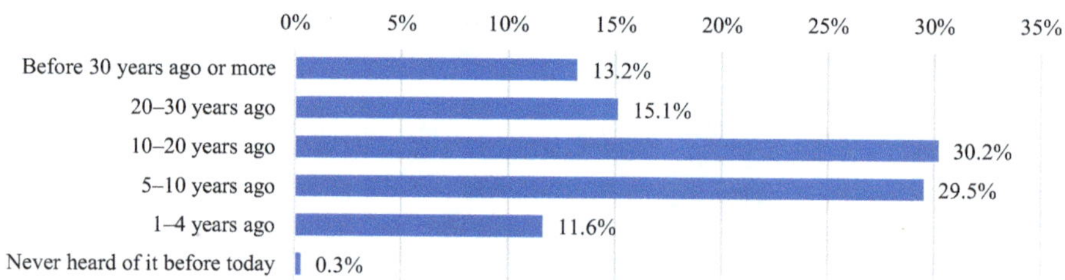

Fig. 3.6 Respondents heard about solar energy for the first time

be recent (skewed to recent years). Most respondents (71.3%) reported learning about solar energy less than 20 years ago, while 15.1% heard about solar energy between 20–30 years ago, and 13.2% answered 30 years ago or more (Fig. 3.6). Based on all the results above, the null hypothesis ($H_0$1) that states, "There is no specific preference of the Saudi population in the source of energy" is rejected and the alternative hypothesis (H1) of specific preference among the Saudi population against using crude oil is accepted.

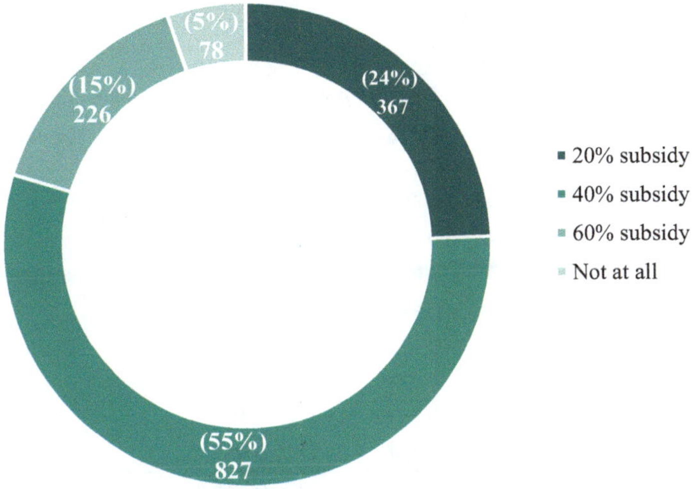

Fig. 3.7 Participants' willingness to install a residential solar photovoltaic system with subsidies

3.5.5 Willingness to Adopt RSPSs

All respondents were directed to answer the question, "Would you consider installing a solar system that costs SAR 7000 and saves you SAR 50 per month if you receive 40% of the system cost as a government subsidy?". The results showed that 79.7% of the respondents (95% CI: 77.7% to 81.7%) were willing to install an RSPS if they would receive 40% back in financial subsidies, while 20.3% (95% CI: 18.3% to 22.3%) rejected RSPS installation with this offer (Fig. 3.7). The acceptors were then asked about their motivations if they would only receive 20% financial subsidies instead of 40%. The majority of the 40% acceptors (69.3%, 95% CI: 66.7% to 71.9%) accepted the suggested 20% financial subsidy, while 30.7% did not see enough value in purchasing an RSPS if they would only receive 20% back in financial subsidies.

Conversely, respondents who rejected the 40% financial subsidy were offered an additional incentive of a 60% subsidy, but even then, only one-quarter of them accepted this offer (25.7%, 95% CI: 20.8% to 30.6%), and the majority of them (74.3%, 95% CI: 69.4% to 79.2%) rejected the offer.

3.5.6 Multivariate Analysis

In the multivariate logistic regression analysis, only two of the study variables were significantly associated with the participants' decisions to accept RSPSs (Table 3.6). These were the frequency of checking electricity consumption and the preferred approach for addressing global warming. The bivariate statistical significance that

Table 3.6 Binary logistic regression analysis for variables with statistical significance in the bivariate analysis ($n = 907$, $R^2_{Nagelkerke} = 0.463$)

Variables	Bivariate p-value	Regression p-value	Effect
Occupation (Ref = others)	0.178	0.124	Confounded
Student	0.144	0.997	
Government employee	0.897	0.997	
Private sector employee	0.643	0.997	
Self-employed	**0.032**[a]	0.996	
Retired	0.882	0.997	
Dwelling Situation (Ref = other)	0.134	0.128	Confounded
Self-built	**0.023**[a]	0.964	
Purchased	0.365	0.445	
Mortgage	0.384	0.806	
Renting	0.051	0.204	
Frequency of checking electricity consumption (Ref = Not at all)	**0.004**[b]	**0.027**[a]	**True Association**
Weekly/Biweekly	**0.001**[b]	**0.000**[b]	
Monthly	0.330	0.080	
Quarterly	0.386	0.470	
Twice a year	0.125	**0.037**[a]	
Annually	0.675	0.999	
Very infrequently	0.250	0.770	
"Global warming is a result of emissions from the use of fossil fuels such as coal, and diesel." (Ref = Strongly Disagree)	**0.017**[a]	0.588	Confounded
Strongly Agree	**0.006**[b]	0.513	
Agree	0.905	0.718	
Neutral	**0.013**[a]	0.541	
Disagree	0.475	0.681	
"Many of the claims about global warming and its impact on the planet are exaggerated." (Ref = Strongly Disagree)	**0.025**[a]	0.406	Confounded
Strongly Agree	0.460	0.995	
Agree	0.322	0.995	
Neutral	**0.025**[a]	0.995	
Disagree	**0.033**[a]	0.996	

(continued)

Table 3.6 (continued)

Variables	Bivariate p-value	Regression p-value	Effect
What do you think is an appropriate approach to address global warming as it is related to energy production? (Ref = Don't know)	**0.000**[b]	**0.003**[b]	**True Association**
Do nothing. We can live with global warming	**0.000**[b]	**0.001**[b]	
Expand nuclear power	0.892	0.346	
Reduce electricity consumption	0.475	0.156	
Expand renewable power	**0.000**[b]	0.724	
Invest in research and development	0.103	0.119	
Do nothing. There is no threat of global warming	0.066	0.101	
If it solved global warming, would you be willing to pay ___ more per month on your electricity bill? (SAR) (Ref = 70 or greater)	**0.010**[a]	0.558	Confounded
No, I prefer to pay nothing	**0.000**[b]	0.577	
10 or less	0.515	0.587	
11–29	0.102	0.996	
30–49	0.117	0.996	
50–69	0.202	0.997	

[a]Statistically significant results (p-value < 0.05)
[b]Highly significant (p-value < 0.01)

was observed in occupation, accommodation status, knowledge of the cause of global warming, knowledge of global warming effects, and the willingness to pay extra money to fight global warming disappeared after controlling for the factor of the checking of electricity consumption and preferences in addressing global warming. This indicates that the last two variables explain the variation observed in the bivariate analysis of all other variables, and therefore have higher predictive power for RSPS rejecters. The results of this question do not consistent with the null hypothesis (H_02) which postulates, "The Saudi population's willingness to adopt an RSPS does not change whether or not they receive subsidies equivalent to 40% of the upfront cost." Thus, we accept the alternative hypothesis (H2) of "The majority of the Saudi population is willing to install RSPS if they would receive subsidies equivalent to 40% of the total cost."

The regression analysis model showed that only two variables were associated with better prediction of the participants' decisions toward RSPS acceptance. Both the frequency of checking electricity and participants' perception toward the appropriate approach against global warming can help explain 26% of the variability in RSPS acceptance ($R^2_{Nagelkerke} = 0.259$). Similar to the bivariate analysis results, participants who checked electricity consumption weekly and biweekly have a 24 times increase in the odds of rejecting RSPS, and participants who do not recommend any measures against global warming have 35 times increase in the odds for rejecting RSPS (Table 3.7).

Table 3.7 Binary logistic regression analysis ($n = 907$, $R^2_{Nagelkerke} = 0.259$)

Variables	B coefficients	Standard Error	Exp (B)	Degrees of Freedom	Regression *p*-value
Frequency of checking electricity consumption (Ref = Not at all)				**6**	**0.010**[a]
Weekly/Biweekly	3.8	1.1	24.2	**1**	**0.000**[b]
Monthly	1.3	0.7	4.0	1	**0.036**[a]
Quarterly	0.9	1.3	2.1	1	0.526
Twice a year	2.4	1.1	14.1	**1**	**0.007**[a]
Annually	−15.8	12139.9	0.000001	1	0.999
Very infrequently	0.3	1.0	3.2	1	0.131
What do you think is an appropriate approach to address global warming as it is related to energy production? (Ref = Don't know)				**6**	**0.001**[b]
Do nothing. We can live with global warming	5.1	1.6	34.5	**1**	**0.003**[b]
Expand nuclear power	1.6	1.7	1.3	1	0.860
Reduce electricity consumption	2.1	1.5	2.8	1	0.424
Expand renewable power	0.5	1.3	0.5	1	0.346
Invest in research and development	2.0	1.3	2.9	1	0.243
Do nothing. There is no threat of global warming	2.6	1.6	4.6	1	**0.001**[b]

[a]Statistically significant results (*p*-value < 0.05)
[b]Highly significant (*p*-value < 0.01)

3.6 Discussion

The demography of study participants generated by the web-based survey does not fully represent Saudi Arabia residents. The study population included 95% of Saudi respondents, while they represent only 62% of the residents in Saudi Arabia. Similarly, the proportion of males who participated in the study was 74% compared to the 57.5% present in the Saudi population as a whole. This reflects the decision-making capacity of the residents, where males and Saudis appear to have more influence on family decisions. Further, the age distribution of the study was different than for Saudi Arabia residents as a whole because, in the study, 68% of respondents were less than 40 years old compared to 46.3% of the Saudi population (GAS 2018). However, all demographic variables were not significantly associated with RSPS acceptance. Therefore, the probability of bias is low.

With no significant variation among the five regions targeted in the present study, the RSPS acceptance rate was considered high compared to other similar studies. While in this study 97% of the respondents were accepting of RSPSs, lower rates were recorded in other studies, such as Yemen (67%), Thailand (67.8%), Malaysia (80%), and Bangladesh (78%) (Baharoon et al. 2016; Poonnavich and Weerin 2015; Solangi et al. 2015; Tareq 2017). This dissimilarity between these results may be partially attributed to the difference in the solar potential and other contextual differences of these countries. The overwhelming acceptance of RSPSs found in the present study analogs with the global trend in energy transition. It also might be in line with the findings of similar studies conducted in the Saudi context (Alrashed and Asif 2015; Dilek et al. 2015; Mosly and Makki 2018), in which participants showed more interest in the installation of PV solar systems as compared to other renewable energy applications. Although the dichotomous question, “Do you accept or reject the notion of installing an RSPS in your dwelling?”, was seeking to understand to what extent do the residents in Saudi Arabia accept or reject the notion of installing an RSPS in their dwelling, the fact that the overwhelming majority was accepting of this notion is very encouraging.

Out of the survey sample, 71.9% of the participants do not prefer crude oil to continue as the main source of energy in Saudi Arabia, supporting the results found among the residents of Dammam city, where 85% see a need to shift from fossil fuel to other resources (Dilek et al. 2015). On the contrary, the public preference of the source of energy used in houses in Turkey was for natural gas (62.5%) over other energy options, including solar energy (21%), even with the assumption of that all the energy options are similar in terms of cost and ease of accessibility. The research group attributed this preference to ease of use, established usage habits, and the perception that natural gas is a clean energy source (Ediger et al. 2018).

In our study, 65% of the participants were in favor of expanding the use of renewable energy to address global warming impact. A similar finding was reported in Yemen, where over 70% of the survey participants agreed that the use of solar energy contributes to protecting the environment and mitigate the climate change effect (Baharoon et al. 2016). The overwhelming preference for shifting from fossil fuel to renewable energy might be explained by the individuals' perception of the consequences of fossil fuels. ISSP's analysis reported that 36% and 26% of respondents of more than 30 countries believe that climate change is very dangerous or extremely dangerous (ISSP 2010). The same study suggested that 16% of the participants definitely believe that fossil fuel is the cause of climate change. In our study, 44% and 31% agree or strongly agree that global warming is a result of the use of fossil fuel. This may indicate the influence of being a resident of a state that is considered the world's largest oil exporter and the world's third-largest oil producer.

While many study variables were significantly associated with RSPS acceptance in the bivariate analysis, only two variables were significantly associated with the acceptance of RSPSs in the multivariate regression analysis. These two variables were first: the monitoring of electricity consumption, and second: the respondent's preferences in addressing global warming. These two variables also explain the differences observed in the bivariate analysis and therefore act as confounders for the other variables. A significance was observed between self-built homeowners in favor of RSPSs which can be attributed to a critical evaluation of the cost-benefit analysis that is reflected in the frequency of electricity consumption monitoring. This is also applicable to self-employed participants who were more likely to conduct a cost-benefit analysis. While income is considered one of the good indicators to assess the economic condition of the respondents, the critical evaluation of the cost-benefit analysis can also provide valuable insights into their economic condition. This was seen in the slightly lower acceptance rate among those who frequently monitor their electricity consumption, as they are more likely to conduct a cost-benefit analysis of RSPSs.

Furthermore, the association between income and monitoring frequency showed highly statistically significant differences ($p = 0.002$). Consequently, some important data are captured by electricity checking frequency, yet not with income. For instance, Table 3.5 shows that participants who check electricity bills weekly or biweekly are likely to have an average monthly income ranging from SAR 5000–19000 (USD 1300–5000), yet not less than SAR 5000 (USD 1300). Those with monthly income less than SAR 5000 (USD 1300) are likely to check their electricity bills monthly, or not checking them at all. This supports the survey results that showed a statistically significant association between RSPSs acceptance and electricity checking frequency ($p = 0.006$), but not with income ($p = 0.747$).

Additionally, the significance of understanding the causes and effects of global warming can be attributed to the preference in addressing global warming. Similarly, the preference for addressing global warming explains the variability in the participants' willingness to pay extra money to avoid the effects of this issue. This indicates that a critical evaluation of the revenue associated with RSPS installation is a major determinant of RSPS rejection, which is consistent with the previous results observed with respect to residence status.

The preference for addressing global warming is congruent with the findings reported in Greece (Ntanos et al. 2018). While findings reported by Ediger et al. (2018), Mosly and Makki (2018), Sommerfeld et al. (2017) showed that economic motivators were seen as the most significant factors associated with purchasing RSPSs.

The results from the multivariate regression analysis can be used to identify the high-risk groups for rejecting RSPSs. The small proportion of participants rejecting RSPSs (3%) were very much influenced by their evaluation of electricity consumption, and these respondents also shared a common negative or apathetic attitude toward global warming. Valuable monitoring strategies may be attractive to these participants. In addition, increasing background knowledge regarding climate change and addressing the effects of global warming may have a significant influence on these participants' decisions regarding solar energy.

Further, more insights could be gained into the respondents' willingness to adopt RSPSs from their answers to the questions of the last section of the survey where 79.7% were willing to install an RSPS provided it would save them 50 SAR/month if they would receive a financial subsidy from the government equivalent to 40% of the total upfront cost to install the RSPS. The majority of those acceptors (69.3%) were also accepting of the proposal, even if only a 20% financial subsidy was offered. While 25.7% of those who rejected the initial 40% financial subsidy were accepting of the idea when a higher subsidy of a 60% was offered, it can be argued that the provision of direct financial support equivalent to 40% or more of the RSPS cost would be a cost-effective motive for encouraging individuals to adopt this technology.

These findings are found to be similar to the results of the public's views analysis in Bahrain (Alsabbagh 2019), where 74% were willing to install RSPSs with subsidy equivalent to 50% of the capital cost. However, the results found in Malaysia are not in line with these findings, whereas 81% of the survey participants were not willing to invest in solar energy through the FiT scheme (Muhammad-Sukki et al. 2011). It is worth mentioning that the GCC countries share many social and economic characteristics, where fossil fuel contributes nearly one-third to total GDP and three-fourths to annual government revenues and exports (IMF 2016). Similar to Saudi Arabia, electricity in the GCC countries, which is mainly generated from oil and natural gas, is consumed largely by the residential sectors of these countries. The Saudi residential sector's share of electricity consumption is 50% (Tlili 2015), a proportion comparable with neighboring monarchies–47% in Bahrain, 33% in Qatar, 64% in Kuwait, 50.2% in Oman, and 39% in the UAE (Alsabbagh 2019; Gastli et al. 2013; Hajiah 2006; Mohaghegh 2015; Taleb and Alsaleh 2014).

These findings are very interesting because they could be correlated to the Saudi political system where an implied social contract between the citizens and government has granted low-priced energy in exchange for political acceptance for decades. Similar to other oil-rich GCC states, cheap energy prices in Saudi Arabia have contributed to the stability of the country for decades, helping to develop a high standard of living and welfare (Kumetat 2012; Ratikainen 2017). The expansion of the renewable energy industry in Saudi Arabia is an area that the government is carefully considering. It has begun to fix the distortions of its domestic economy by abolishing power source subsidies. For example, in January 2018, electricity tariffs for the residential sector were raised by 80–260% (Al Eqtisadiah 2017). These developments need to be followed closely in order to further assess the impacts on the adoption rate of RSPSs.

It has been clear that the respondents well recognize the challenges posed by air pollution and global warming. They showed a preference to escape from fossil fuel as the main source of energy in the country to other alternatives where they see renewable energy as an optimal option. As the results revealed, the supportive perception of protecting the environment and the positive association with solar energy can serve as motives for personal adoption, if an appropriate incentive scheme is introduced rather than the Net Metering scheme that is currently introduced.

The recent increasing trend in electricity tariffs helps RSPSs to compete against conventional energy sources. However, further hikes could be applied in order to enhance the economic viability of RSPSs, taking into consideration the necessity to readjust the financial support programs dedicated to low and middle-class citizens (e.g., Citizen Account Program) in order to alleviate the pressure caused by ongoing economic reforms. Political buy-in for renewable energy is essential, yet this should be translated into providing financial incentives to induce the adoption of solar PV systems at the household level. Although many countries have chosen to phase out incentive schemes (e.g., Feed-in tariff) or limit pay for incentivized adoption of RSPSs, it is important to consider the rationale behind these actions because those countries have either achieved or exceeded their set goals for diversified energy. In other cases, such countries have taken measures to prevent the grid fluctuation caused by oversupply from RSPSs. Conversely, in a country like Saudi Arabia where RSPSs were recently introduced, and only 1.45% of households currently use solar PV systems (GAS 2019) it is imperative to introduce incentivizing mechanisms to stimulate the dissemination and mass adoption of these technologies.

3.7 Recommendations

Every endeavor to expand the adoption base of RSPSs will not be effective unless the households are aware of the advantages and disadvantages of these applications. Therefore, awareness campaigns should be collectively carried out by the pertinent government and semi-government agencies (including REPDO, K.A.CARE, ECRA, and SEC) along with concerned businesses and universities. A well-informed household is often in a better position to make sound decisions regarding the adoption of RSPSs. Further, developing a national interactive platform where individuals can evaluate the cost efficiency of their RSPS purchase decision by assessing both inputs and outputs of such an investment (i.e., calculating the required installation space, estimated initial cost, estimated surplus electricity, and the RSPS payback period) can assist them in making a decision based on relevant figures. Another aspect that could be of interest is investigating the electricity consumption behavior of households before and after the adoption of RSPSs. Understanding not only if their consumption behavior has changed or not, but also to what extent it might have changed, could support the government's effort to achieve its domestic energy efficiency and diversity objectives.

3.8 Limitations

This study has a few limitations which need to be addressed in order to improve future research. Missing data occurred for two questions. However, the analysis of the other variables indicated that these were Missing Completely at Random (MCAR) because they were not significantly associated with any other variable. The MCAR notation usually does not cause significant changes or bias in the results of the survey questions. These missing data could have resulted from an unknown system error in the online survey form, supporting their classification as MCAR. In addition, the study was influenced by an online survey methodology, which was reflected in different demographic characteristics, even though controlling for these demographics indicated no significant variation in their subcategories.

In this study, institutional and corporate levels were not investigated, but rather this research was primarily focused on exploring the residents' views toward RSPSs in Saudi Arabia. Therefore, further wide-ranging research could be of interest as it would help to elaborate on the complexity of the solar energy landscape in Saudi Arabia. Such a comprehensive study would also help the government to work proactively, along with other stakeholders, to identify and overcome barriers to the entire solar PV value chain blocking the dissemination of RSPSs.

3.9 Conclusion

In this study, the public's views on the adoption of RSPSs in Saudi Arabia were explored. A particular focus was provided to basic demographic characteristics, residence status, household energy use, environmental and energy cognition, and willingness to adopt RSPSs. Results show that using solar energy as an alternative source of energy is, by large, a preferable option among the respondents, with no significant variation observed in acceptance or rejection rates among different regions, age groups, or income brackets. This indicates a significantly high level of social acceptance in Saudi Arabia compared to other countries. It also reflects the strong desire of individuals in Saudi Arabia to expand the utilization of solar energy on both a personal and state level.

While the vast majority of the respondents recognized the potential causes and effects of global warming and believed that expanding the usage of solar energy could address this challenge, most respondents were resistant to paying more than SAR 10 (USD 2.5) for their monthly electricity bill to mitigate this issue. Further, a majority of participants in this study agreed to purchase an RSPS under the currently introduced scheme of Net Metering, if they could receive a government subsidy equivalent to 40% of the RSPS's upfront cost. The study also identified that the most significant determinants for social acceptance of RSPSs in Saudi Arabia were a critical evaluation of electricity consumption and participants' preferences in addressing global warming. Saudi residents recognize the necessity to shift from the reliance on oil exportation to solar energy and other RETs, which was reflected in their interest in installing RSPSs. Besides the fresh insights, this study revealed—resulting in benefits to local communities, policy-makers and other stakeholders in the solar energy market—the generalization of these findings to other oil-exporting countries are guided by similarities in context, specifically low-priced electricity tariffs and per capita income, which are common characteristics shared by other GCC states.

References

Abdul Hai, M. (2019). Rethinking the social acceptance of solar energy: Exploring "states of willingness" in Finland. *Energy Research & Social Science, 51*, 96–106. https://doi.org/10.1016/j.erss.2018.12.013.

ADA [WWW Document]. (2018). Arriyadh Development Authority. Retrieved 28, July 2019, from http://rda.gov.sa/ADA_e/DocumentShow_e/?url=/res/ADA/En/Projects/RiyadhMetro/index.html.

Al Eqtisadiah [WWW Document]. (2017). 80 to 260% increase in electricity tariff for residential sector. Retrieved 9, July 2019, from http://www.aleqt.com/2017/12/13/article_1298286.html.

Al Kobraish, H. (2018). Personal communication with Haytham M. Al Kobraish, CEO, ALKOBRAISH INVESTMENT & CONSTRUCTION, at Qassim First Energy Forum, Al Bukayriyah, Al Qassim Region, 2018.04.12.

Al-Barrak, S. (2018). Personal communication with Shareef Al-Barrak, the Vice Director Technical Affairs of Electricity & Cogeneration Regulatory Authority ECRA, at the headquarters located in Riyadh, 2018.04.25.

Aldraehim, M. S., Edwards, S. L., Watson, J., & Chan, T. (2013). Cultural Impact on e-service use in Saudi Arabia: The need for Service Oriented Culture. Presented at the 19th American Conference on Information Systems, Chicago, Illinois, USA, p. 9.

Almusabihi, A. (2018). *Al-Falih: Increasing energy prices is indispensable.* Okaz newspaper.

Alrashed, F., & Asif, M. (2015). An exploratory of residents' views towards applying renewable energy systems in saudi dwellings. *Energy Procedia, 75,* 1341–1347. https://doi.org/10.1016/j.egypro.2015.07.207.

Alsabbagh, M. (2019). Public perception toward residential solar panels in Bahrain. *Energy Reports, 5,* 253–261. https://doi.org/10.1016/j.egyr.2019.02.002.

AL-Shehry, A. (2006). The Motivations for Change towards e-Government Adoption 22.

Alshehry, A. S., & Belloumi, M. (2015). Energy consumption, carbon dioxide emissions and economic growth: The case of Saudi Arabia. *Renewable and Sustainable Energy Reviews, 41,* 237–247.

Al-Taher, M. (2018). Personal communication with Mohammed Y. *Al-Taher, the Distribution Planning Department Manager of Saudi Electricity Company SEC, at the headquarters located in Riyadh, 2018*(05), 14.

Arthur D Little [WWW Document]. (2018). Saudi Arabia: World leader in solar power? Retrieved 21, Aug 2019, from https://www.adlittle.com/en/insights/viewpoints/saudi-arabia-world-leader-solar-power.

Ayaz, K., Akbar, G., & Aslam, A. (2003). World Wide Fund for Nature (Eds.), 2003. Conservation and Islam, 1st ed. World Wide Fund for Nature, Islamabad.

Baharoon, D. A., Rahman, H. A., & Fadhl, S. O. (2016). Publics' knowledge, attitudes and behavioral toward the use of solar energy in Yemen power sector. *Renewable and Sustainable Energy Reviews, 60,* 498–515. https://doi.org/10.1016/j.rser.2015.12.110.

Bryman, A. (2012). *Social research methods* (4th ed.). Oxford, New York: Oxford University Press.

Chaplin, C. (2014). Imagining the Land of the Two Holy Mosques: The Social and Doctrinal Importance of Saudi Arabia in Indonesian Salafi Discourse. *Austrian Journal of South-East Asian Studies, 7,* Volumes per year: 1 p. https://doi.org/10.14764/10.aseas-2014.2-6.

Claudy, M., Driscoll, A. O.', & Duffy, A. (2010). Home Owners' Attitudes, Perceptions and Willingness to Pay for Microgeneration Technologies. https://doi.org/10.13140/2.1.3269.2483.

Clayton, V. (2012). KSA leads the world in mobile usage | | ITP.net [WWW Document]. Retrieved 29, Jan 2020, from https://www.itp.net/588729-ksa-leads-the-world-in-mobile-usage.

DeVellis, R. F. (2016). *Scale development: Theory and applications.* New York, The United State of America: SAGE Publications Inc.

Dilek, D., Nahid, S., Noor, F., & Deemah, A.-Q. (2015). Public acceptance of renewable energy and Smart-Grid in Saudi Arabia | Request PDF, in: ResearchGate. Presented at the Proceedings of the 8th IEEE GCC Conference and Exhibition, Muscat, Oman. https://doi.org/10.1109/IEEEGCC.2015.7060018.

Ediger, V. Ş., Kirkil, G., Çelebi, E., Ucal, M., & Kentmen-Çin, Ç. (2018). Turkish public preferences for energy. *Energy Policy, 120,* 492–502. https://doi.org/10.1016/j.enpol.2018.05.043.

Edison, S. W., & Geissler, G. L. (2003). Measuring attitudes towards general technology: Antecedents, hypotheses and scale development. *Journal of Targeting Measurement and Analysis for Marketing, 12,* 137–156. https://doi.org/10.1057/palgrave.jt.5740104.

Electricity & Cogeneration Regulatory Authority ECRA (2017). Small-Scale Solar PV Systems Regulations.

Erhun, K. (2001). Islam and environmental conservation. *Environmental Conservation, 28,* 1–9. https://doi.org/10.1017/S0376892901000017.

GAS (The total population of Saudi Arabia). (2018). *General Authority of Statistics.*

Gastli, A., Charabi, Y., Alammari, R., & Ali, M. (2013). Correlation between climate data and maximum electricity demand in Qatar. Presented at the 2013 7th IEEE GCC Conference and Exhibition (GCC), IEEE, Doha, Qatar. https://doi.org/10.1109/IEEEGCC.2013.6705841.

Hajiah, A. E. (2006). Energy Conservation Program in Kuwait: A Local Perspective, in: Proceedings of the Fifteenth Symposium on Improving Building Systems in Hot and Humid Climates, ESL-HH-06-07-47. Orlando.

Huijts, N. M. A., Molin, E. J. E., & Steg, L. (2012). Psychological factors influencing sustainable energy technology acceptance: A review-based comprehensive framework. *Renewable and Sustainable Energy Reviews, 16,* 525–531. https://doi.org/10.1016/j.rser.2011.08.018.

Ilieva, J., Baron, S., & Healey, N. M. (2002). Online Surveys in Marketing Research. *International Journal of Market Research, 44,* 1–14. https://doi.org/10.1177/147078530204400303.

IMF. (2016). International Monetary Fund, Economic Diversification in Oil-Exporting Arab Countries. Policy Papers 16. https://doi.org/10.5089/9781498345699.007.

International Social Survey Programme Research Group. International Social Survey Programme: Environment III—ISSP, 2010.

IRENA. (2013). Renewable energy innovation policy: Success criteria and strategies. *The International Renewable Energy Agency.*

Kumetat, D. (2012). *Managing the transition: An analysis of renewable energy policies in resource-rich arab states with a comparative focus on the United Arab Emirates and Algeria.* London: The London School of Economics and Political Science.

MobileGate. Study: Saudi Arabia ranks first in the list of countries most used for smartphones in the world [WWW Document]. (2017). Retrieved 22, Aug 2019, from https://bit.ly/2MocNeG.

Mohaghegh, S. (2015). Modelling material use within the low carbon energy pathways by using TIMES model generato.

Mosly, I., & Makki, A. (2018). Current status and willingness to adopt renewable energy technologies in Saudi Arabia. *Sustainability, 10,* 4269. https://doi.org/10.3390/su10114269.

Muhammad-Sukki, F., Ramirez-Iniguez, R., Abu-Bakar, S. H., McMeekin, S. G., & Stewart, B. G. (2011). An evaluation of the installation of solar photovoltaic in residential houses in Malaysia: Past, present, and future. *Energy Policy, Clean Cooking Fuels and Technologies in Developing Economies, 39,* 7975–7987. https://doi.org/10.1016/j.enpol.2011.09.052.

Ntanos, S., Kyriakopoulos, G., Chalikias, M., Arabatzis, G., & Skordoulis, M. (2018). Public perceptions and willingness to pay for renewable energy: A case study from Greece. *Sustainability, 10,* 687. https://doi.org/10.3390/su10030687.

Oxford Dictionaries [WWW Document]. (2012). Definition of acceptance in English. https://www.lexico.com/definition/acceptance.

Poonnavich, S., & Weerin, W. (2015). Critical Factors of Social Acceptance for Solar PV Rooftop in Thailand. Presented at the 2015 4th International Conference on Informatics, Environment, Energy and Applications, Pattaya, Thailand. https://doi.org/10.7763/IPCBEE.

Ratikainen, K. W. (2017). *Transitioning to renewable energy in Saudi Arabia.* A multi-level perspective analysis of the Saudi renewable energy policies: University of Oslo, Norway.

REN21 [WWW Document]. (2017). Renewables 2017 Global Status Report. Retrieved 27, July 2019, from http://www.ren21.net/gsr-2017/.

RenewablesNow [WWW Document]. (2016). Renewablesnow.com. Retrieved 10, Oct 2019, from www.news/saudi-arabia-unveils-plan-for-50-mw-solar-project-486513.

Saudi Vision [WWW Document]. (2016a). Saudi Vision 2030, Media Center. Retrieved 23, July 2019, from https://vision2030.gov.sa/en/media-center.

Saudi Vision [WWW Document]. (2016b). Retrieved 30, Jan 2020, from https://vision2030.gov.sa/en/node/87.

Solangi, K. H., Saidur, R., Luhur, M. R., Aman, M. M., Badarudin, A., Kazi, S. N., et al. (2015). Social acceptance of solar energy in Malaysia: users' perspective. *Clean Techn Environ Policy, 17,* 1975–1986. https://doi.org/10.1007/s10098-015-0920-2.

Sommerfeld, J., Buys, L., & Vine, D. (2017). Residential consumers' experiences in the adoption and use of solar PV. *Energy Policy, 105,* 10–16. https://doi.org/10.1016/j.enpol.2017.02.021.

Sütterlin, B., & Siegrist, M. (2017). Public acceptance of renewable energy technologies from an abstract versus concrete perspective and the positive imagery of solar power. *Energy Policy, 106,* 356–366. https://doi.org/10.1016/j.enpol.2017.03.061.

Taleb, H., & Alsaleh, Y. (2014). Applying energy-efficient water heating practices on the residential buildings of the United Arab Emirates. *International Journal of Environmental Sustainability, 9,* 35–51. https://doi.org/10.18848/2325-1077/CGP/v09i02/55084.

Tareq, F., 2017. Acceptance and willingness to pay for pv solar electricity: survey evidence from southern part of Bangladesh. Presented at the International Conference on Mechanical Engineering and Renewable Energy, Chittagong, Bangladesh, p. 7.

Tlili, I. (2015). Renewable energy in Saudi Arabia: Current status and future potentials. *Environment, Development and Sustainability, 17,* 859–886. https://doi.org/10.1007/s10668-014-9579-9.

Trading Economics [WWW Document]. (2018). Saudi Arabia GDP per capita. https://tradingeconomics.com/saudi-arabia/gdp-per-capita.

T-Systems [WWW Document]. (2018). NEOM: Saudi-Arabia plans to build the world's most high-tech city. Retrieved 28, July 2019, from https://www.t-systems.com/en/best-practice/01-2018/fokus/smart-world/neom-782556.

Visschers, V. H. M., & Siegrist, M. (2014). Find the differences and the similarities: Relating perceived benefits, perceived costs and protected values to acceptance of five energy technologies. *Journal of Environmental Psychology, 40,* 117–130. https://doi.org/10.1016/j.jenvp.2014.05.007.

Woertz, E. (2013). *The domestic challenges in the saudi energy market and their regional and geopolitical implications.* Norway: The Norwegian Peacebuilding Resource Centre.

Chapter 4
Critical Factors Influencing Social Perception of Residential Solar Photovoltaic Systems in the Hydrocarbon-Rich Saudi Arabia

Building on the date demonstrated in Chap. 3, in this chapter, a study is carried out in a descriptive style to get an in-depth understanding of influencing determinants on the adoption of RSPSs in Saudi Arabia. The second part of the questionnaire used herein comprises three sections including the demographic status section, solar energy cognition section, and an innovation attributes assessment section. Using *DOI* theory as a framework for analysis, critical factors of accepting or rejecting the notion of adopting RSPSs are assessed on the Rogers' innovation attributes (*relative advantage*, *observability*, *complexity*, *compatibility*).

4.1 Research Background

Both uncertainties over energy supply and the worldwide rapid population growth increase the tension and motivation to secure energy. These threats are becoming more severe as the effects of climate change become more harmful. Therefore, a shift toward renewable energy generation has been a global trend. For many countries of the Middle East and North Africa, in general, and the member states of the GCC, in particular, substantial revenues derived from oil and natural gas exports have been an essential element in the development and growth of these countries. This was reflected in maintaining high levels of infrastructures and welfare, besides expanding regional and global influence through large-scale investments and sovereign wealth funds. This particularly holds true for Saudi Arabia, which has one of the world's largest proven oil reserves. Taking a leading role in oil production and exporting for decades, allowed Saudi Arabia to become a great power influencing the global energy landscape. However, in the past few years, Saudi Arabia has taken significant steps to add renewable energy into its energy mix, seeking a sustainable and clean energy supply. Different ministries and commissions have been working jointly under

K. A. Alrashoud, *Managing Technology Transition in Saudi Arabia*,
https://doi.org/10.1007/978-981-15-5803-0_4

the newly announced economic reform plan Saudi Vision 2030, launching several initiatives and projects to achieve this goal (Saudi Vision 2016).

Solar energy is seen as a promising alternative source of energy; harnessing it could take advantage of the high direct normal irradiance ranging between 9000 watt-hours per square meter per day (Wh/m^2/day) in summer months and 5000 Wh/m^2/day in winter months (AlYahya and Irfan 2016b). KACARE was established by the government to promote the integration of indigenous renewable energy resources by conducting research and providing necessary data (KACARE 2010). KACARE worked KACST along with other government agencies and universities to develop a roadmap for the successful development of the renewable energy industry. The roadmap was modified several times before it was comprehensively redesigned. In 2017, the MOE established the REPDO to deliver renewable energy across Saudi Arabia. REPDO's primary mission is to ensure that the goals of renewable energy projects align with Saudi Vision 2030 and to ensure that all stakeholders involved in energy generation, regulation, measurement, research, and data acquisition are performing in harmony to achieve the countries' renewable energy goals (REPDO 2017).

In 2018, Saudi Arabia began developing renewable energy capability upward of 150–200 (GW) that would span the entire value chain, from manufacturing to local and regional project development (Arthur D Little 2018; Bridge 2019). Saudi Arabia has also taken steps recently to encourage individuals to generate their electricity by adopting RSPSs. Since July 2018, individuals in Saudi Arabia have been granted the right to install solar PV systems at their homes and connect them to the public grid (ECRA 2017; SEC 2015). This measure is meant to restrain the surging electricity consumption in the residential sector, which accounts for approximately 50% of the total electricity consumed locally (Tlili 2015). Saudi Electricity Company (SEC), the local electricity company, has called upon individuals wishing to install RSPSs at their residence to submit their applications. However, only a small number of applications (approximately 3000 as of May 2018) have been received, which falls short of expectations (Al-Taher 2018). The utilization of the RSPSs can be beneficial to individuals and relieve the public grid from electricity shortage; yet, the adoption of RSPSs in Saudi Arabia still encounters hurdles that need to be addressed. There are a number of studies reviewing the current situation in the context of renewable energy technologies in Saudi Arabia. However, these studies mainly focus on reviewing the current situation of solar energy landscapes (AlYahya and Irfan 2016a; Ratikainen 2017; Salam and Khan 2018) comparing several types of renewable energy technologies (Hepbasli and Alsuhaibani 2011; Mosly and Makki 2018; Ramli et al. 2016), or different aspects (technical-focused analysis) of the renewable energy status in Saudi Arabia (Al-Ghamdi and Alshaibani 2018; Almasoud and Gandayh 2015; Asif 2016; Dehwah et al. 2018). These studies indicated the high potential in Saudi Arabia for renewable energy in general, and for solar energy in particular. The residential sector was also indicated as a promising area to capitalize for RSPSs, yet the requisite need to address the existing gap on social-related studies of solar energy in Saudi Arabia was also highlighted (Dehwah et al. 2018; Ratikainen 2017). In Saudi Arabia, there has been no in-depth official or academic investigation of the diffusion of RSPSs from a social perspective based on an innovation-oriented theoretical framework (Al-Barrak 2018; Al-Taher 2018). Given the important role that the residential sector can play in generating electricity in Saudi Arabia together with the lack of an in-depth, social-related investigation toward solar energy acceptance, it is vital to acquire robust and reliable data that elaborates on how Saudi residents perceive the adoption of RSPSs (Al-Barrak 2018; Al-Taher 2018).

As stated in Chap. 3, the author of this book are aware of some articles in which authors explicitly differentiate between the terms "acceptance" and "acceptability," where the former indicates a supportive behavior of an object or situation, and the latter refers to an attitude toward an object or saturation (Huijts et al. 2012). However, in this book, the author defines the word "acceptance" to be "agreement with or belief in an idea or explanation (Oxford Dictionaries 2012) by showing a favorable attitude or positive reception." Based on the *DOI* theory, this study sheds light on the factors that may be associated with the acceptance or rejection of RSPSs by individuals in Saudi Arabia. *DOI* theory was used, among other theories, as a framework to conduct this research due to its applicability to the Saudi context and its close relationship between the system of social structure and the diffusion of innovations (Rogers 1983).[1] Therefore, this theory was used to conceptualize and develop an empirical investigation rather than to suggest a modification to the theory itself. Since *DOI* theory is a critical per-condition theory of how, why, and at what rate new innovations are spread, any novel endeavor to examine and illustrate the systemic process by which an innovation diffusion occurs is valuable, in both the theoretical and policy-making domains.

In the study of (Poonnavich and Weerin 2015), which the analysis of the chapter was drawn from, the public views of adopting RSPSs in Thailand were investigated yet with a slightly different framework and approach. The present study employed a study sample of a larger size. It also adopted a different approach and conducted a more sophisticated data analysis. Besides the scientific contribution that this study seeks to make towards the field of research, the novelty of both the context and policy implications will redound to the benefit of the local community, considering the significant role that the residential sector could play in the energy transitions currently underway in Saudi Arabia.

Furthermore, energy initiatives in Saudi Arabia can diffuse easily into another similar Gulf Cooperation Council Arab states. In addition, as one of the world's largest fossil fuel suppliers, energy transitions in Saudi Arabia could have a significant influence on energy choices internationally. The present study, therefore, will provide fresh insights into these areas by studying prospective adopters of RSPSs in Saudi Arabia.

[1]Professor Everett Rogers's book of Diffusion of Innovations theory was first published in 1962, and currently the book is now in its fifth edition (Rogers 2003).

4.2 Diffusion of Innovations Framework

4.2.1 The Mechanism of Innovation Diffusion

The concept of innovation diffusion has received considerable attention over the years. The *DOI* theory was originated by the communication theorist and sociologist Everett Rogers (Rogers 1983). Rogers defines an innovation as "an idea, practice, or object that is perceived as new by an individual or other unit of adoption." Other sociologists define innovation variously, similarly when categorizing the process of innovation adoption. Robertson broadens the definition by referring to innovation as a "process whereby a new thought, behavior, or thing is conceived of and brought into reality" (Robertson 1967). While Schumpeter narrows the definition of innovation as "setting up of a new production function" (Schumpeter 1939). Barnett, on the other hand, defines innovations as "any thought, behavior, or thing that is new because it is qualitatively different from existing forms" (Barnett 1953).

Rogers describes diffusion as a process by which innovation is communicated over time among participants of a social system through certain channels (Rogers 1983). Rogers suggests that four elements have a major impact on the diffusion of innovations: the innovation itself, social system, timing, and communication channels. Innovation diffusion is highly subject to the adopters' characteristics. Based on the degree to which adopters accept a new idea, diffusion manifests itself in various forms (Rogers 1983).

In this context, *DOI* theory categorizes adopters into four groups: early adopters, early majority, late majority, and laggards (Rogers 1983). Generally, adopters go through five primary stages until they make the actual action to adopt a new idea. These stages are knowledge, persuasion, decision, implementation, and confirmation (Rogers 1995). The knowledge stage commences when the individual realizes the existence of the innovation for the first time and starts to formulate knowledge about its nature and how it functions. In the third stage, persuasion, the individual forms an attitude toward the new idea. This attitude might be favorable or unfavorable, depending on the individual's action based on their characteristics. The term persuasion herein is oriented toward the receiver rather than toward the source. The next stage in the innovation-decision process is the decision stage, which is formed when a receiver engages in activities that result in deciding to adopt or reject the innovation. The fourth stage is the implementation stage, where an individual puts innovation into use. All three preceding stages, knowledge, persuasion, and implementation are mental exercises. However, in the implementation stage, these thoughts are summed up and put into practice, thus involving overt behavior change. The last stage in the innovation-decision process is the confirmation stage, at which an individual seeks reinforcement for the decision he or she has already made. At the confirmation stage, the individual tends to avoid any cause of dissonance and reduces it whenever it occurs (Rogers 1995).

The diffusion process is multidimensional and varies based on the type of innovation (Karakaya et al. 2014). The adoption of innovations like solar PV systems is not necessarily determined by its availability in the market. Instead, it is subject to how adopters perceive the innovation attributes, which explains the wide variance (49–87%) of diffusion rates of different new ideas (Rogers 1983; Tidd 2010). Rogers identified a comprehensive set of innovation attributes. These attributes represent the persuasion stage of Rogers' model for an "innovation-decision process", and it includes *relative advantage*, *observability*, *complexity*, *compatibility*, and *trialability* (Rogers 1983).

Based on the *DOI* theory, the *relative advantage* is defined as "the degree to which an innovation is perceived as being better than the idea it supersedes" (Rogers 1983). To ensure successful diffusion, an innovation must be perceived as offering greater advantages when compared to that of existing services or products. These advantages can be demonstrated in economic profitability, effort saving, impact on comfort, or an improvement in any other manner upon existing products or practices. *Observability* is "the degree to which the results of an innovation are visible to others" (Rogers 1983). Innovation can be adopted faster if it is visible to others. Generally, the results of some ideas are easier to be described and observed than describing the nature of innovation. Therefore, the *observability* attribute is considered to be a critical component in improving the diffusion effect. The third attribute, *complexity* (simplicity), is "the degree to which an innovation is perceived as relatively difficult to understand and use" (Rogers 1983). The easier it is to use the innovation, the faster it diffuses, and vice versa. Some innovations are clear in their meaning to prospects adopters, while others require clarification to be understood. The fourth attribute is *compatibility,* which is defined as "the degree to which an innovation is perceived as consistent with the existing values, past experiences, and needs of potential adopters" (Rogers 1983). Innovation must be in line with social norms and not cause a conflict with existing services or products. Some innovations require more time before they become acceptable. The fifth attribute, trialability, is "the degree to which an innovation may be experimented with on a limited basis" (Rogers 1983). This includes innovations that can be temporarily tried and dispensed, making them easier to adopt after the trial. While this option is not common for most innovations (e.g., solar PV systems), wide adoption of a particular innovation can reduce the uncertainty a potential adopter might have before he or she decides to adopt it (Rogers 1983).

Much attention has been given to the consumers' perspective toward the adoption of renewable energy technologies over the past 20 years. Moreover, their consumption behavior is seen as a critical factor since it exceeds ecological limits, and its effect on the depletion of natural resources has become a highly discussed topic (Byers 1996). In general, consumers tend to resist to green innovation. Although the advantages of innovative services or products may overweigh that of traditional ones, they fail to penetrate the market or acquire a considerable market share (Faiers et al. 2007). According to (Faiers et al. 2007) 40–90% of innovations across different product categories fail to survive in the market, and consumer resistance is ascribed as the primary cause of this failure. Hence, the dilemma regarding the adoption of small-scale solar PV systems must be investigated because of its critical role in electricity consumption behavior in Saudi Arabia. Every society has various types of adopters with different cultural backgrounds and different demographic characteristics. This variation leads to different levels of speed in the adoption of innovation (Neame 2009).

4.2.2 Consumer Behavior and Innovation Diffusion

A considerable number of studies have focused on consumer behavior in the innovation of diffusion; however, studies considering the diffusion of solar energy from the consumer behavior perspective have appeared only recently (Faiers and Neame 2006; Leonard-Barton 1980). The decision by a household to adopt an RSPS might be affected by various factors. These factors include endogenous elements (e.g., desire to conserve energy and technology awareness) and exogenous elements (e.g., market system, cost, and characteristics of PV systems) (Qureshi et al. 2017).

An exploratory survey conducted by Claudy et al. (2010) among the individuals in the Republic of Ireland; showed that factors such as upfront cost, the complexity of the solar PV system and the interference of the solar PV system with the day-to-day living style were critical. Another study carried out in Germany by Jacobsson and Lauber (2006) suggested that the adoption of an RSPS as a concept is socially accepted by virtue of its environmental value. Yet, economic feasibility is the main determinate of a successful adoption.

A cost-benefit study on various options of microgeneration including solar energy, micro-wind, and micro combined heat and water concluded that even with the existence of incentive mechanisms, the long payback periods contributed to the failure of producing electricity production through microgeneration in the United Kingdom (Allen et al. 2008). In Thailand, (Poonnavich and Weerin 2015) analyzed the social acceptance factors involved in the adoption of RSPSs. In their study, a total of 400 responses were collected through a survey conducted among individuals. Approximately 65.7% of the participants accepted the solar PV rooftop systems. Relying on the analysis of the collected data, they suggested that respondents with a higher level of education have a stronger tendency to adopt RSPS in their home. Poonnavich and Weerin also argue that the sense of contributing to solving the global warming situation by using renewable energy sources seems to be the most significant factor that encourages the respondents to purchase an RSPS. In developing countries, a low level of technology awareness among potential adopters was identified as a critical hurdle. Thus, misleading information disseminated among social actors about technological advantages, besides the phobia of high cost, can prevent the diffusion of technologies (Muntasser et al. 2000; Wamukonya 2003). In Pakistan, a study identified unawareness within the local community, high upfront cost, inadequate renewable energy policy, and lack of technical expertise as barriers in the diffusion of solar PV utilization (Khan and Latif 2010). Similarly, another study conducted in India suggested that economic and financial factors are the biggest hurdles in widespread solar PV adoption at the household level (Eltawil and Zhao 2010). In comparison, in Nigeria, component failure, inadequate policies, building incompatibility, and complications in research and development were found to be the most significant barriers (Akinboro et al. 2012).

In the Middle East, many studies have investigated the potential of solar energy in terms of technical and legislative aspects. However, studies addressing social-related aspects have only rarely been conducted (see Sect. 1.2 for literature in the Saudi context). Regardless of the enormous potential, the Middle East states have, the integration of solar energy is still in the early stages (Seznec 2018).

The rapid changes in the landscape of energy supply and frequent failure or unavailability of the supply of electricity in remote areas of different parts in Saudi Arabia put more pressure on houses in urban areas to install RSPSs. This situation may ultimately boost the diffusion of RSPSs in the local context.

4.3 Research Questions and Hypotheses

4.3.1 Research Questions

The research in this chapter has two goals. The first goal is to elucidate, from Saudi residents' perspective, the level of acceptance or rejection of the RSPSs. The second goal is to investigate the most significant factors that act as determinants of the individuals' decision by examining the following questions:

RQ2. To what extent do individuals in Saudi Arabia accept the notion of installing a residential solar PV system?

RQ3. What are the most critical factors influencing the individuals' decision to accept or reject the adoption of a residential solar PV system in Saudi Arabia?

4.3.2 Research Hypotheses

The following hypotheses were developed and tested by using empirical data to find the relationship between the factors included within the scope of this work.

H3. The majority of the Saudi population accepts the notion of installing a residential solar PV system in their homes.

H4. The most significant deciding factor in accepting residential solar PV systems in Saudi Arabia is the *revenue* obtained by reducing electricity consumption and selling excess electricity to the service provider.

4.4 Methodology and Research Design

Based on an extensive review of the extant literature on the adoption of renewable energy technologies, the author attempted to select the most suitable approach to achieve the study objectives. The present study carried out in a descriptive and exploratory style to get an in-depth understanding of influencing determinants on the adoption of RSPSs in Saudi Arabia by utilizing a case study approach. A case study is an empirical inquiry that examines a phenomenon within its real-life context. It is based on an in-depth investigation to gain insights and explore the causes of underlying principles which subsequently can guide future inference to similar contexts as well as identifying inference limitations (Yin 2009). Therefore, an in-depth study, of different attributes of Rogers' *DOI* theory, using the case study approach, was conducted to provide a better understanding of potential determinants influencing the adoption of solar PV systems among households in the hydrocarbon-rich state, Saudi Arabia. The study was a cross-sectional web-based survey of respondents in the five main areas in Saudi Arabia (Central, Northern, Eastern, Southern, and Western areas). This design was more suitable as the primary research question discusses the extent of RSPSs acceptance, which is best described through frequencies and proportions. Compared with other longitudinal study designs, the cross-sectional study presents the advantage of assessing the prevalence, which accurately reflects the extent of RSPS acceptance (LEE 1994).

A cross-sectional web-based questionnaire was distributed in a convenient way in several online channels, including e-mails, LinkedIn, WhatsApp, Twitter, and Facebook. While the study used a non-probability convenience sampling method that tries to represent the decision-making capacity in the Saudi communities, the author exerted many efforts to guard against selection bias. The survey was distributed to all trending social, educational, political and media groups. Moreover, region-specific groups were also searched, and followers were invited to participate in the survey. In contrast, to avoid bias, the researchers avoided channels such as academic groups associated with renewable energy and environmental protection groups as they are expected to have greater knowledge and interest in solar energy. Considering the large youth segment in the Saudi population (46.3% are under the age of 40) which is also ranked first worldwide in smartphone use, a web-based approached was chosen as it appeals to a much wider audience with respect to cost and time constraints (Clayton 2012; GAS 2018; MobileGate 2017). A pilot sample was used to test the validity of the questionnaire before it was deployed. The questionnaire used in this study comprises a demographic status section, solar energy cognition section, and innovation attributes assessment section.

Data were gathered from March 2018 to May 2018 with 1500 interaction times and two missing attempts, thus resulting in 1498 collected responses; the error rate was <1% at a 95% confidence level. This limits the type I error rate to (0.05), where results with a *p*-value less than 0.05 are described as statistically significant, and results with $p < 0.01$ are described as highly significant.

4.4.1 *Questionnaire Design*

The web-based questionnaire was designed into sections to address the study questions and the postulated hypothesis. It included sections on participants' demographics, solar energy cognition, and innovation attributes assessment section. While the innovation attributes assessment section was directly related to the study questions, those on the demographic and the solar energy conditions are essential to determine the possible sources of bias, confounders, as well as determining gaps in participants' knowledge on the topic. The study used a previously tested questionnaire that was developed based on Rogers's *DOI* theory (Poonnavich and Weerin 2015). This reproduction of the questionnaire was important to allow comparability of study findings with other published papers, and therefore determine the specific characteristics that are related to the Saudi population. The questionnaire was reviewed by two experts in the field under study, and the Arabic translation was also revised and piloted to ensure ease of understanding and comparability. The reliability of the questionnaire was further tested as described in the Data Analysis section.

4.4.1.1 Demographic Status

The respondents were asked questions, and they could select one answer among several options. First, they were required to choose what best matches their demographic status including nationality, gender, residential location, age, occupation, education, and income.

4.4.1.2 Solar Energy Cognition

In the second section, the questionnaire assessed the respondents' perception of solar energy by asking them four questions. These four questions asked for the participants' own opinions on the reliability of solar energy, the negative impact of solar energy on the environment, job creation in the solar power industry, and the negative impacts of solar energy on local communities. The participants' answers were received as a 5-point Likert scale, using the labels (1: Very low), (2: Low), (3: Moderate), (4: High), and (5: Very high), where lower values indicated weaker agreement with the statement, and higher values refer to a stronger agreement with the statement. A Likert scale, also known as Likert-type scale, is a psychometric scale that is widely used for scaling responses in survey research (Likert 1932). The scale was introduced by Rensis Likert in 1932 using a 5-point scale to a statement to which respondents assess their level of agreement or disagreement. Although the 5-point scale might become 7 points or 3 points, and the agree-disagree scale may become high–low, very likely–not likely, the principle of the Likert scale remains the same (Joshi et al. 2015; Likert 1932). In comparison to single statements, the Likert scale was chosen given the potential to increases reliability, especially when measuring attitudes, emotions orientations and intentions (Likert 1932).

4.4.1.3 Innovation Attributes Assessment

Following these questions, the respondents were provided with a brief description of solar PV systems and a simplified diagram illustrating how the system works. The respondents then were directed to answer a dichotomous Yes/No question, "Do you accept or reject the notion of installing an RSPS in your dwelling?" Based on their answers to this question, the respondents were divided into two groups, acceptors and rejectors. Next, each group was directed to answer 20 questions to precisely rate the possible factors that may have influenced their decision to accept or reject the notion of installing an RSPS in their home. The 20 questions were developed based on the *DOI* theory and categorized by four innovation attributes, namely *relative advantage*, *observability*, *complexity*, and *compatibility*. Each attribute represents five factors that respondents were asked to assess on a 5-point Likert scale.

4.4.2 Innovation Attributes and Factors Selection

Due to the complexity of Saudi Arabia's electricity generation landscape and adoption of RSPSs, theoretical frameworks are more appropriate for the case-study investigation compared with technical frameworks. The former extends beyond the technical aspects of solar PV systems diffusion, by including the socio-economic and stakeholders' perception, to give a comprehensive picture of solar PV systems diffusion at the households' level. Considering such a framework is justified because of the interrelated multi-determinants that are involved in the adoption decision. With that being said, as critical per-condition theory, the *DOI* theory was selected for the key role it could play in understanding how an idea or product diffuses through a specific population. Despite criticisms of the *DOI* theory (e.g., the one-way information flow, emphasis on individual blame at the expense of a more nuanced consideration of the influence of the social structure), the *DOI* theory was used in a wide range of fields, including but not limited to, agriculture products, e-services, medical treatment procedures, and renewable energy technologies (Faiers and Neame 2006; Rogers 1983; Solvak et al. 2019). The *DOI* theory is most effective when applied to new product launches, as well as when introducing an existing product to a new market.

Considering an RSPS as an innovation (only 1.45% of houses in the administrative regions of Saudi Arabia use solar PV systems) (GAS 2019), the attributes of the *DOI* theory can provide a valid basis from which to draw conclusions about the determinants that motivate or impede individuals from adopting this technology in Saudi Arabia. There are major challenges to the dissemination of RSPSs, including obstacles that are common to the dissemination of new technologies, as well as those more related to the solar PV systems, such as the electricity grid and the built environment. Although there are several obstacles in the solar PV value chain, the present study seeks to identify critical factors that may act as motivators or disincentives throughout the perspective of sociotechnical systems regarding solar PV systems dissemination in Saudi Arabia.

The data collected were mainly assessed based on Rogers's innovation characteristics (*relative advantage, observability, complexity, compatibility*), which by potential adopters evaluate innovations. Each attribute plays a relative role, either positively or negatively, in the dissemination of RSPSs. Hence, the effect of these attributes was examined by asking the respondents to evaluate a list of factors that were subsumed under these attributes (Table 4.1).

Based on Rogers's characteristics of innovations, these factors include all possible aspects that could influence individuals' decision-making. These factors have often appeared as strong determinants in studies on adopting renewable energy microgeneration technologies. The factors were selected, distributed, and carefully modified to fit into the Saudi context while maintaining the scope of each attribute as suggested by Rogers (1983).

Although Rogers summarized his interpretation of each attribute, there are no absolute rules for what constitutes a certain attribute as one. For instance, regarding *relative advantages*, Rogers suggests that "The degree of *relative advantage* may be measured in economic terms, but social-prestige factors, convenience, and satisfaction are also often important components. It does not matter so much whether an innovation has a great deal of "objective" advantage. What does matter is whether an individual perceives innovation as advantageous" (Rogers 1983). The more positively a *relative advantage* of an innovation is perceived, the more rapidly it is going to be adopted, and vice versa.

While Poonnavich and Weerin limited the factor revenue included under the *relative advantage* attribute as "sell solar electricity" (Poonnavich and Weerin 2015), the present study has broadened the definition of *revenue* as "reduce consumption and sell unused generated electricity". The rationale is that both can fulfill economic advantage; generating revenue (sell excess electricity generated through a PV system to the national grid) and save on utility bills (consume less electricity imported from the national grid by consuming electricity generated from a PV system). The other four factors included under the *relative advantage* attribute (i.e., *environment protection, unlimited power*, *global warming*, and *technology development*) fall within "social-prestige, convenience, and satisfaction" components, as stated earlier (Rogers 1983).

Regarding, *observability* attribute, similar to any other innovation, observable results reduce uncertainty, encourage peer discussion, and increase the knowledge about the innovation, which eventually leads to higher rates of adoption. "Solar panels on a household's roof are highly observable, and a California survey found that the typical solar adopter showed his equipment to about six of his peers" (Rogers and Adhikarya 1979). The visibility of innovation, as perceived by members of a social system, is positively related to its rate of adoption and vice versa (Rogers 1983). This is correlated with the local positive perception (neighbors' attitude),

Table 4.1 Innovation attributes and factors definition

Attributes	Factors	Factors definition
Relative advantage	Revenue	Reduce consumption and sell unused generated electricity
	Environment protection	Reduce Greenhouse Gas emission
	Unlimited power	Limitless power from the sun
	Global warming	Mitigating global warming impact
	Technology development	Solar power technology is expected to be developed constantly in the future
Observability	Income statistics monitoring	Revenues from electricity fed back to the grid can be monitored
	Neighbor attitudes	Neighbor Attitude regards environment protection image
	Power production monitoring	Electricity generation can be monitored
	Greenhouse gas reduction monitoring	Greenhouse Gas emission can be monitored
	Solar energy knowledge	System installation increases the knowledge about it
Complexity	Installation space	Area required for system installation
	Availability of service providers	Difficulty of access to service providers
	Residence location	Solar radiation potential of the location
	Maintenance	Complexity of maintenance
	Residence structure	Durability of the roof and residence structure for system installation
Compatibility	Power system	Effect of a solar system on the existing one
	Land Uuse	Area of the system effect on the living pattern of the installer
	Installation cost	Installation cost effect on living cost
	Global trends	Compatibility of solar energy systems with global trends
	Social values	Compatibility of solar energy systems with social values

Reproduced with permission from (Poonnavich and Weerin 2015)

which is affected by their knowledge of innovation, as well as its functionality monitoring (Rogers 1983). For instance, the more electric vehicles (EVs) appear in the streets, the more likely members of a social system to stimulate peer discussions and develop a piece of knowledge about them and ultimately form a perception or stance toward them. The same applies to RSPSs that are commonly observable on rooftops, which subsequently increase the probability of triggering peer discussions about their functionality (Karakaya 2015).

Other measurable visible results of using a PV system are the amount of electricity generated by the PV system, statistics of revenues generated from electricity fed back to the grid, and monitoring Greenhouse Gas reduction. These factors fall within the visible components of the *observability* attribute and could act as strong determinants.

As for the *complexity* (simplicity) attribute, the simpler an innovation is to understand and use, the more likely it is to be adopted. Innovations that require adopters to develop new skills or understanding are less adopted than those that do not require such an effort (Rogers 1983). From the individuals' perspective, the *complexity*-simplicity in understanding and using an RSPS can be represented by the installation space, maintenance, availability of service providers, and residence structure and residence location. Certainly, there are also various external factors that may contribute to the *complexity*–simplicity continuum of RSPSs. However, in order to achieve the objectives of the present study, this component is restricted to the perspective of individual adopters.

Concerning *compatibility*, it is essential for an innovation to be compatible with potential adopters' needs, social, and religious values for easier adoption. Innovation should also be perceived as consistent, not only with deeply imbedded social norms but also with the past experience and the existing life patterns (Rogers 1983). In a socially and religiously conservative country such as Saudi Arabia, these components could greatly influence social actors' decision-making. This includes the perception of the innovation as a socially-compatible addition, which in turn is affected by a positive image globally recognized as a trending product (Beise 2003; Rogers 1983). In addition, the innovation logistics (installation costs and land use) and integrability with the existing power system make it more compatible to adopt and consequently more favorable (Rogers 1983).

The literature showed that both motives and disincentives determinants are essentially context-dependent, (e.g., electricity tariffs, the economic stability of residents, type of available electricity sources, benefits, and drawbacks of solar PV systems under local scenarios, etc.). Nevertheless, for critical motives and disincentives, most of these conditions and factors can be subsumed under the innovation attributes described in Rogers' *DOI* theory (Rogers 1983; Vasseu and Kemp 2015).

4.4.3 Date Analysis

Reliability testing across the four innovation attributes scales was conducted to check the internal consistency of the acquired set data. Normality testing was done to determine the appropriate statistical methods for data analysis (parametric or nonparametric tests). Shapiro–Wilk test and Kolmogorov–Smirnov tests are recommended in the literature for testing normality as they are superior in power for detecting significant deviations from normality (D'Agostino and Stephens 1986). These tests provide statistically significant results ($p < 0.05$) whenever the distribution of the numerical variable significantly differs from the normal distribution (D'Agostino and Stephens 1986). Nonparametric distributions are best described with the median for central tendency, the interquartile range for dispersion and the Mann-Whitney U test for comparing different medians with statistical significance. Furthermore, Spearman's rho rank correlation coefficient test was calculated to determine the rank correlation between the variables (Lehman et al. 2005). Further, the mean scores of the questions per attribute were calculated along with the frequencies and proportion for each question to identify the important variables in respondents' attitudes. The chi-squared test and Fisher exact test were used to compare the categorical variables. The corresponding p-value was calculated for each statistical test, where results with a p-value of less than 0.05 were considered to be statistically significant, and results with a p-value of less than 0.01 were considered to be highly significant.

4.5 Results

The reliability testing results ranged between 'acceptable' and 'good' (DeVellis 2016), where Cronbach's alpha internal consistency results ranged between 0.735 to 0.814 for different scales in the study. The deletion of any question included in each of the four scales did not significantly improve the internal consistency (less than 5% changes), reflecting a good construct formulation. The results of the reliability tasting across different attributes shown below in Table 4.2.

4.5.1 Demographic Status

Analysis of the respondents' demographic characteristics indicated that the majority were Saudi nationals, male, young (<40 years old), employed, and undergraduate educated, having a monthly income of less than 14,000 Saudi Riyals (SAR) (USD 3.733), as shown in Table 4.3. The examination of the collected data shows no statistically significant variation in the acceptance or rejection rates among all demographic variables except for occupation ($p = 0.015$), where the self-employed category had the lowest acceptance rate (94.7%) in comparison to all other categories. Moreover,

Table 4.2 Reliability testing across different attributes

Attribute	Number of iems	Cronbach's alpha				If question deleted	
		Accept RSPS	Reject RSPS	Overall	Internal consistency	Question	Cronbach's alpha value
Relative Advantage	5	0.750	0.759	0.773	Acceptable	RA1	0.81
Observability	5	0.812	0.866	0.814	Good	OB1	0.809
Complexity	5	0.809	0.719	0.812	Good	CX3	0.788
Compatibility	5	0.722	0.772	0.735	Acceptable	CB3	0.739

Table 4.3 Frequencies and proportions of residential solar photovoltaic systems acceptance based on demographic characteristics ($n = 1498$)

Characteristic		Frequency	Percent (%)	Accept RSPS (%)	*p*-value (with RSPS Acceptance)
Nationality	Saudi	1420	94.8	97.5	1.000[b]
	Non-Saudi	78	5.2	97.4	
Gender	Males	1139	73.8	97.1	0.114
	Females	359	26.2	98.6	
Area	Central area	744	49.7	97.6	0.105[b]
	Eastern area	232	15.5	97.0	
	Southern area	96	6.4	96.9	
	Western area	353	23.6	98.6	
	Northern area	73	4.9	93.2	
Age	18–30	495	33.0	98.8	0.101[b]
	31–40	531	35.4	96.8	
	41–50	259	17.3	96.1	
	51–60	167	11.1	98.2	
	61–70	43	2.9	95.3	
	71 or more	3	0.2	100	
Occupation	Student	244	16.3	99.6	0.015[a, b]
	Government employee	524	35.0	97.1	
	Private sector employee	339	22.6	96.8	
	Self-employed	131	8.7	94.7	
	Retired	152	10.1	97.4	
	Other	108	7.2	100.0	
Education	Postgraduate	274	18.3	98.9	0.067[b]
	Undergraduate	874	58.3	96.7	
	Secondary or equivalent	314	21.0	98.7	
	Junior high school	29	1.9	93.1	
	Elementary school	6	0.4	100.0	
	Informal education	1	0.1	0	
	Not gone to school at all	1	0.1	100.0	
Income (SAR)	Less than 5000	117	7.8	96.6	0.747[b]

(continued)

Table 4.3 (continued)

Characteristic		Frequency	Percent (%)	Accept RSPS (%)	p-value (with RSPS Acceptance)
	5000–9000	302	20.2	97.4	
	10,000–14,000	350	23.4	96.3	
	15,000–19,000	242	16.2	97.9	
	20,000–24,000	192	12.8	97.4	
	25,000–29,000	85	5.7	98.8	
	30,000–34,000	63	4.2	100.0	
	35,000–39,000	41	2.7	97.6	
	40,000 or more	106	7.1	99.1	

[a]Statistically significant results ($p < 0.05$)
[b]For these characteristics, Fisher exact test was used since some of the expected values were lower than 5, making chi-squared test results less accurate

as summarized in Table 4.3, no significant variation in the acceptance or rejection rates was observed among the five targeted areas in this study (Central, Eastern, Southern, Western, and Northern areas).

4.5.2 *Solar Energy Cognition*

As can be seen in (Fig. 4.1), the general trend in the data shows a good knowledge of solar energy. Most participants agree on the continuity and persistence of solar

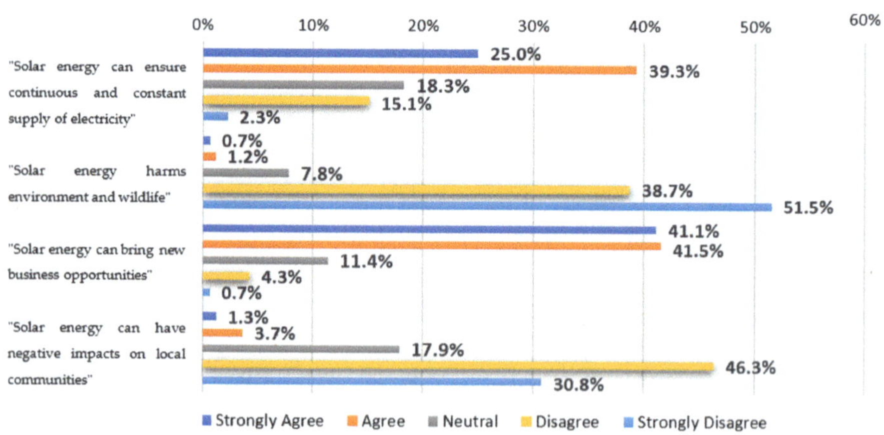

Fig. 4.1 Respondents knowledge of solar energy ($n = 1498$)

energy-based electricity supply, disagree on solar energy harmfully affecting the environment and wildlife, agree that solar energy brings new business opportunities, and disagree on negative impacts on local communities.

The respondents were asked to assess four general statements based on a 5-point Likert scale. A total of 64.3% of participants agree (or strongly agree) with the statement "Solar energy can ensure a continuous and constant supply of electricity," while 17.4%) disagree and 18.3% have a neutral attitude. While the responses to the statement "Solar energy harms environment and wildlife" were more one-sided with 90.2% not agreeing with the statement, only 7.8% of the respondents choose a neutral opinion. Moreover, the vast majority (82.6%) agree with the statement, "Solar energy can bring new business opportunities." On the other hand, 5% disagree and 11.4% have an indifferent position.

Furthermore, rejection of the statement "Solar energy can have negative impacts on local communities" predominates (77.1%), while a neutral stance of 17.9% and a significant low agreement rate of 5% was observed (Fig. 4.1).

4.5.3 Innovation Attributes Assessment

When the respondents were asked whether they accept or reject the notion of installing an RSPS, the overwhelming majority (97.46%) accepted installing an RSPS at their residence (95% CI: 96.7% to 98.3%). Therefore, these results are consistent with the postulated hypothesis (H3), which states the "The majority of the Saudi population accept the notion of installing a residential solar PV system in their homes."

Regarding the results of the four attributes, the scores given to each attribute show that acceptors preferred to agree on the *relative advantage* attribute (median score 22), whereas rejectors preferred to agree on the *observability* attribute which was given the highest score in comparison to other attributes (median score 15). See Table 4.4.

Concerning the *relative advantage* attribute, most of the acceptors agree on the positive effect s of all variables of the *relative advantage.* The *unlimited power* variable, which refers to the inexhaustibility of solar energy for the generation of electricity, had the highest rates of respondents who agree on its positive effects. On the other hand, respondents who reject RSPS installation had higher disagree rates on the *revenue* component of solar energy, as shown in (Fig. 4.2a and Table 4.5).

Regarding *observability*, the results demonstrated a general trend for RSPS acceptors to agree on the positive effect s of income monitoring, i.e., power production monitoring and solar energy knowledge, where the highest of these was income statistics monitoring. In contrast to all other variables across different attributes, RSPS rejectors had a higher rate of agreement on the neighbor attitudes variable as compared to that of RSPS acceptors (Fig. 4.2b and Table 4.5).

The analysis of the *complexity* attribute revealed a general trend that RSPS acceptors agree on the positive effect s of all *complexity* variables except for the availability of service providers. The highest agree rate of these variables was for the

Table 4.4 Statistics across different attributes and tests to ascertain differences in medians

Attribute	Testing normality		Median	Interquartile Range	Median by RSPS acceptance			*p*-value (Independent Samples Mann-Whitney U Test)
	Kolmogorov-Smirnov	Shapiro-Wilk			Accept RSPS	Reject RSPS	Median Difference	
Relative Advantage	<0.001**	<0.001**	22	6	22	13	9	<0.001**
Observability	<0.001**	<0.001**	18	7	18	15	3	<0.001**
Complexity	<0.001**	<0.001**	18	7	18	12	6	<0.001**
Compatibility	<0.001**	<0.001**	19	6	20	14	6	<0.001**

**Highly significant ($p < 0.01$)

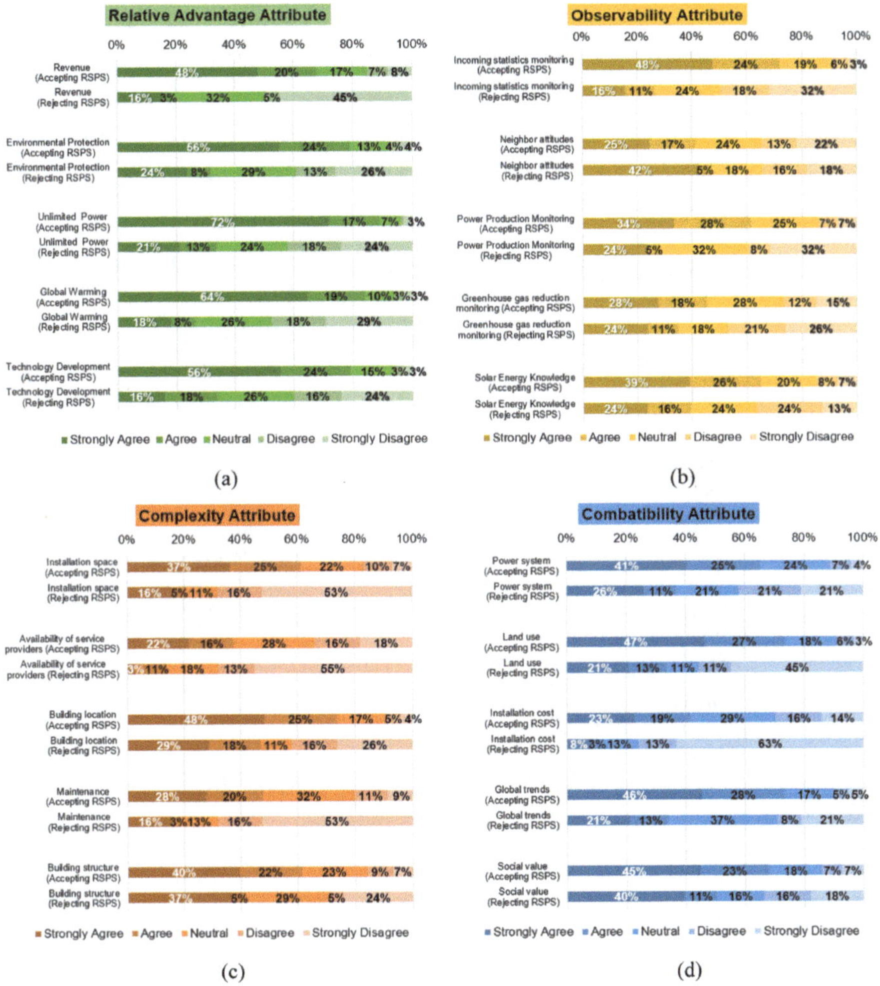

Fig. 4.2 Participant response regarding the variables of the innovation attributes. **a** Participant response on *relative advantage* variables, **b** participant response on *observability* variables, **c** participants response on *complexity* variables, **d** participants response on *compatibility* variables (In order to make the comparison between acceptors and rejectors easier, the questions in the rejectors Likert scale were recorded in reverse order to match the acceptors' scale)

residence location. On the other hand, respondents who rejected RSPSs installation show several trends in disagreeing on the positive effect s of installation space, availability of service providers (which had the highest disagree rate), and maintenance (Fig. 4.2c and Table 4.5).

Concerning *compatibility*, RSPS acceptors generally agree on the positive effects of all variables related to the *compatibility* attribute, except for the installation cost, which scored 3.20 as the Likert scale mean value (Fig. 4.2d and Table 4.5). The highest

Table 4.5 The association between attributes/variables and accepting/rejecting RSPS installation

Attribute	Variable	Mean of Likert Scale		Mean difference	The average mean difference for attribute
		Accept RSPS	Reject RSPS		
Relative advantage	Revenue	3.93	2.39	1.54	1.53
	Environment Protection	4.24	2.89	1.35	
	Unlimited Power	4.55	2.89	1.65	
	Global Warming	4.38	2.68	1.69	
	Technology Development	4.27	2.87	1.40	
Observability	Income Statistics Monitoring	4.08	2.61	1.47	0.66
	Neighbor Attitudes	3.10	3.37	−0.27	
	Power Production Monitoring	3.74	2.82	0.93	
	Greenhouse Gas Reduction Monitoring	3.32	2.84	0.48	
	Solar Energy Knowledge	3.82	3.13	0.69	
Complexity	Installation space	3.74	2.16	1.59	1.13
	Availability of Service Providers	3.10	1.92	1.17	
	Residence Location	4.09	3.08	1.01	
	Maintenance	3.46	2.13	1.33	
	Residence Structure	3.80	3.26	0.54	
Compatibility	Power System	3.91	3.00	0.91	1.08
	Land Use	4.08	2.55	1.53	
	Installation Cost	3.20	1.79	1.41	
	Global Trends	4.06	3.05	1.00	
	Social Values	3.92	3.37	0.55	

among these values was land use with a 4.08 mean value. Those who rejected RSPS installation had the highest disagree rate on the compatibility of installation cost.

The association between attributes and RSPS acceptance shows that the *relative advantage* was the most critical determinant since the difference in the median score between acceptors and rejectors was 9 points (Table 4.4). Additionally, the most significant variable among all attributes was the effect on *global warming*, where the mean among the acceptors was 1.69 points more than that among the rejectors (Table 4.5). The second most significant variable was *unlimited power*, and the third was *revenue*. These results are not consistent with the postulated hypothesis (H4) which states, "The most significant deciding factor in accepting residential solar PV systems in Saudi Arabia is the *revenue* obtained by reducing electricity consumption and selling excess electricity to the service provider." Income statistics monitoring, installation space, and land use had the highest mean difference among *observability*, *complexity*, and *compatibility* variables, respectively (Table 4.5).

None of the four attributes scores were normally distributed, as the statistical significance of Kolmogorov–Smirnov, and Shapiro–Wilk shows in Table 4.4 (therefore, nonparametric statistics are used for description and analysis, i.e., median, interquartile ranges, and Spearman's rho correlation coefficient). The highest median score was seen in the *relative advantage* attribute, which is also consistent with the higher median difference between RSPS accepting and rejecting groups, thus indicating a stronger association with the decision to accept RSPS installation. The lowest median difference was seen in the *observability* attribute.

The results of the correlation analysis between the four attributes indicate a moderate positive correlation (0.50–0.70), except for correlations between the *relative advantage* and *complexity* as well as the correlation between *observability* and *compatibility*, which were a low positive correlation (0.30–0.50). Furthermore, the strongest association was seen between the *relative advantage* and *observability* scores, followed by the correlation between *complexity* and *compatibility* scores[2] (Table 4.6).

4.6 Discussion

Many countries are reconsidering energy strategies and utilizing renewable energy. Social acceptance is a key factor in the success or failure of the sought-after energy transition. It has been argued that institutions (policy and market) depend on the cognition of social actors (socio-cognitive learning), where institutions are perceived as "inter-subjective" social reality (Geels 2010). The outcomes of public review surveys form the basis of regulations and policies developed to promote the adoption

[2]The results show that the distribution of attributes scores was rightly skewed for participants accepting RSPSs installation in all four attributed, where the figures showed more skewness in the *relative advantage* attribute. On the other hand, the graphs for those who reject RSPSs installation did not show any specific trends, except for the complexity score which was skewed to the left toward lowest scores (disagree on the positive effects) (see Annex i).

Table 4.6 Correlation between different scales used in the study

		Relative advantage score	Observability score	Complexity score	Compatibility score
Spearman's rho Correlation Coefficient	Relative Advantage Score		0.633**	0.466**	0.507**
	Observability Score	0.633**		0.530**	0.487**
	Complexity Score	0.466**	0.530**		0628**
	Compatibility Score	0.507**	0.487**	0.628**	

**Correlation is significant at the 0.01 level (2-tailed)

of specific energy technologies. Therefore, it is crucial to identify how to evaluate acceptance in public review studies to ensure that the outcomes of these studies provide accurate and representative data that allow valid inferences to be made about local acceptance. Additionally, debates on energy supply infrastructure are often irrational and emotionally driven, as individuals at some level hold certain perceptions about different types of energy technologies. To allow for knowledgeable and informed decision making, it is essential to have a solid understanding of the psychological aspects of the representation of alternative energy systems and how to effectively assess them.

The content presented in this book shows that there are various paths toward expanding the use of solar PV systems in Saudi Arabia. Each individual participating in this study followed particular motivations, implicating a specific set of identities and competencies. The demographic characteristics of the study participants who were collected by the web-survey do not fully represent the population in Saudi Arabia. The study population had 95% Saudi national participants, whereas they represent 62% of the population in Saudi Arabia. Similarly, the gender of the participants (74% of the study being male as compared to males making up 57.5% of the population) and the age distribution (68% were less than 40 years in comparison to 46.3% of the population being under 40) were not representative of all Saudi Arabia residents (GAS 2018). Nonetheless, no significant association with the RSPS was observed among these demographic variables; therefore, the probability of bias remains low. Furthermore, the acquired study population reflects the decision-making capacity in a masculine society, where males in Saudi Arabia have more influence on family decisions (Alsubaie and Jones 2017).

The rate of accepting the notion of installing RSPS found in this study answers the (RQ2) and is seen to be remarkably high when compared with the results of other similar studies used the same quantitative research approach. While Poonnavich and Weerin indicated that the acceptance ratio in Thailand was 67.8%, Tareq reported the number of 78% in the southern part of Bangladesh (Tareq 2017). Similarly, two separate studies in the northern part of Pakistan and southern Punjab revealed that 81% and 82% of the respondents, respectively, were interested in RSPSs (Abdullah et al. 2017; Kamran et al. 2019). A study in Malaysia contended that 80% of the participants in the survey were highly interested in RSPSs (Solangi et al. 2015)

This dissimilarity may be attributed to the different circumstances of the studies. The significantly positive image and high rate of acceptance of RSPSs found in this study may also be in line with a similar study conducted among residents in Saudi Arabia, in which the participants' preferences for solar PV systems outweighed other given options of renewable energy applications (Alrashed and Asif 2015; Mosly and Makki 2018). The respondents' preference for solar energy may have been directly affected by the increasing number of announcements about large utility-scale solar energy projects in different parts across Saudi Arabia, which are being carried out under the national plan Saudi Vision 2030 that aims for a less oil-centered and more sustainable economy. Additionally, the long solar irradiation time Saudi Arabia enjoys and the rapid solar energy development in the neighboring countries (e.g., The United Arab Emirates and Jordan) may have also affected their perception.

Regardless of the variation of the development levels in the infrastructure, education system, and lifestyles across the country, the results show no significant variation among respondents from all five targeted regions of the country. These findings may point to no cultural-based differences in perception regarding solar energy technology across all regions in Saudi Arabia and can likely be attributed to the popularity of solar energy-based power, where electricity is only partially or completely not available (e.g., remote farms and desert pasture).

The median score of the *relative advantage* attribute was observed as the highest among the other attributes. These values are consistent with the results of Spearman's rho correlation coefficient test. Among the factors examined in the study, the location and structure of a house clearly act as determinants in the acceptance or rejection of RSPSs. Moreover, among all factors of the four innovation attributes, *global warming*, ranks as the most significant variable with a mean difference score of 1.69, followed by *unlimited power* (1.65) and *installation space* (1.59) (Table 4.5). While *unlimited power* is the most agreed upon variable, it also has a slightly higher agreed rate among rejectors. Therefore, the lack of appreciation of global warming among rejectors contributed more to their negative view of RSPSs. These prominent findings answer the research question (RQ3) and reject the postulated hypothesis of (H4) that suggests the factor *revenue* is the most significant deciding factor in accepting RSPSs. Although the variation between the scores of the three factors, *global warming*, *unlimited power*, and *installation space*, is low, the *global warming* factor ranking first might be attributed to the large young segment of respondents with current knowledge on global warming and safe renewable power sources, thus reflecting the population structure in Saudi Arabia. The level of awareness of global warming found in this study is consistent with that of Al-Mutairi's study (Al-Mutairi et al. 2019), which assessed the level of awareness and knowledge about climate change among the community of the rural region of Tabuk (northern Saudi Arabia); and the study concluded that individuals involved in the study showed a moderate level of knowledge toward climate change issues, impacts, and solutions proposed by the authors. It is noteworthy that the *availability of service providers* had the lowest mean of Likert scale (3.10) among acceptors compared to other factors included in the *complicity* attribute. This finding may contradict the findings of a similar study conducted in Lahore, Pakistan by Qureshi et al. (2017), where the majority of respondents (94%) agreed with the easy availability of these systems in the market. Besides the narrower gap between the cost of electricity generated by conventional means and solar energy in Pakistan, this finding might be also attributed to the dissimilarity of the infrastructure advancement levels, where more areas with intermittent access to electricity exist in Lahore.

The highly positive attitude toward RSPSs installation is an encouraging finding for the promotion of these technologies among individuals and for achieving the desired energy transition in Saudi Arabia. These findings also cause concern because they do not reflect the small number of applications made to install such technology (Al-Taher 2018), which indicates an existing gap between the individuals and the pertinent government agencies. Calling for application submission without building a solid foundation of policies and knowledge could cause a significant variation in the perception of prospects. This has important implications for the underlying policies, regulations, and information provision for RSPSs in Saudi Arabia. To allow for intact decision making, it is vital to educate the targeted segment (i.e., households) by providing them with accurate information about solar energy technologies.

4.7 Recommendations

Based on the findings of this study, the author recommends developing a national interactive platform, where interested individuals can assess the required upfront investment, estimated electricity output, required space, and potential revenue based on their dwelling location prior to the purchase of RSPSs; they should be given the opportunity to develop a piece of knowledge in advance about both the advantages and disadvantages of RSPSs which consequently will help to ensure a successful and timely implementation of the technology.

Moreover, the Net Metering mechanism currently introduced by the Electricity & Co-Generation Regulatory Authority (ECRA 2017), might not seem attractive to end-users, as it only allows any excess electricity generated by end-users to be banked to their account for future consumption. Taking into consideration the high amount of electricity consumed by air conditioners in Saudi Arabia (especially in desert climate areas), it is less likely that any excess electricity will be accumulated for further consumption. Thus, an alternative incentive mechanism must be introduced to the public in a way that fosters plausible and constant injection of solar harvested electricity into the grid. Besides the guaranteed uptake, the author suggests that the current Net Metering mechanism should be modified to allow individuals to sell back any excess electricity and receive payments based on how valuable the exported electricity is to the electricity system at certain times. (i.e., greater payment should be received for electricity exported in peak times). This would encourage individuals to reduce their electricity consumption during peak times by trying to secure excess electricity to export to the grid to take advantage of the high uptake rates at these particular times. Since the temperature and levels of energy demand vary significantly during the year in Saudi Arabia, the SEC has built a number of high-cost electricity generation plants across the country that are only to be used for a very short period of time (summer peak times) to reduce energy supply failure caused by the high demand for electricity. This challenge might be overcome if a time-based uptake scheme is introduced for homeowners.

Additionally, despite the fact that many countries have phased out feed-in tariff schemes with no replacement, and despite the consistently reduced cost for solar PV system components, it is essential in the Saudi context to design mid-term subsidizing programs that relieve the high upfront cost of RSPSs and encourage plausible integration without causing fluctuation or disrupting the standability of the electricity system. It is imperative for a country like Saudi Arabia, which consumes the equivalent of 1.5 to 2 million barrels of oil daily in electricity generation (Aliqtisadi 2016), to develop a national direct subsidy plan dedicated to homeowners to increase the RSPSs adoption rates. The increasing local consumption of a substantial portion of the total liquid petroleum is cutting directly into the oil export volume, which could have otherwise been exported at global market rates. Giving the fact that the petroleum is sold to the SEC at as low as 3% of the international market prices (approximately USD 3 per barrel) (Okaz 2015) and considering that the total government subsidies spent on energy and water in 2015 reached nearly 79 billion dollars (Reuters 2017), it could be argued that a portion of the government energy subsidies should be allocated for the support of homeowners in purchasing RSPSs. Similar to other GCC countries, in an absolute monarchy like Saudi Arabia, revenues derived from exported oil and gas has facilitated to what so-called "ruling bargain", a mainly top-down social contract in which governments provide a wide range of subsidized services in return for their citizens' political acceptance (CIRS 2013). The provision of low-priced energy has been, for decades, a primary element by which GCC states achieved their social, economic, and political objectives (APICORP 2016). This justifies the present image of the survey respondents, where the non-financial benefits (limiting *global warming* and enabling *unlimited power* sources) are more valuable compared to financial benefits (*revenue*). However, this policy has come at huge cost. Besides the fact that substantial portion of crude oil is being consumed domestically, which prevents Saudi Arabia from maximizing the value of its natural resources and extending its export base, low energy prices have created an illusion of abundance, which consequently resulted in excessive increase in energy consumption coupled with rapid population growth (APICORP 2016; Ratikainen 2017).

It is fair to mention that Saudi Arabia has recently taken major steps in reforming the government subsidy structure, following the steps of the other GCC states. It has been gradually abolishing subsidies applied to various utility services including electricity. For instance, starting from January 2018, electricity tariffs for the residential sector have been hiked between 80–260% (Al Eqtisadiah 2017). Although the new tariffs are still low according to international and regional standards, such measures would make purchasing an RSPS economically feasible as it helps to shorten the solar PV systems payback period and allow them to compete against hydrocarbons sources.

Furthermore, the study recommends that the government should consider adopting a zero-rating tax on solar PV system components in a way that does not violate the international trade law or other agreements that Saudi Arabia is a part of. Abolishing tax would be of assistance as it reduces the upfront cost of purchasing an RSPS, thus making it more affordable for individuals to possess. In addition to the recommendation above, an introduction of long-term, low-interest loans can encourage more extensive implementation of RSPS among individuals. Unguaranteed high-interest loans may not fit conceptually and can be counterproductive as they increase the overall cost of solar PV systems by adding the cost of margin interest payments.

4.8 Limitations

The empirical results reported herein should be considered in light of some potential limitations that need to be addressed to improve future research. The survey was conducted online, which was reflected in various demographic characteristics, even though controlling for these demographics indicated no significant variations in their subcategories. Additionally, the selected online survey methodology was not applicable to those who do not have access to the internet. Moreover, the questions used in the questionnaire were closed-ended, Likert type scale questions and a dichotomous question. Although these types of questions provide precise and clear answers, they do not provide a specific explanation of why a particular answer was chosen among other options. However, regardless of how the respondents formed their knowledge about solar energy and how they used the information presented in the questionnaire, the findings of this study clearly provide a valid basis from which to draw conclusions about the social perception of RSPSs adoption in Saudi Arabia.

4.9 Conclusion

The purpose of this research was to investigate the social perception of RSPSs in hydrocarbon-rich Saudi Arabia. A cross-sectional, web-based survey was conducted among individuals to examine what factors may potentially act as a motivator or disincentive in the adoption of RSPSs based on the *DOI* theory. The findings of this study demonstrated the individuals' overwhelming positive perception of the RSPSs with no significant variation in the acceptance or rejection rates among participants from the five main geographic regions of Saudi Arabia included within the scope of this work.

While the self-employed participants had a lower acceptance rate for the installation of RSPS, there were no significant differences between educational and income categories. Participants' knowledge of solar energy was good, where the majority had sound views of solar energy renewability, environmental and communities' safety, and the creation of new job opportunities. The study also reviewed participants' perceptions of attributes and factors related to the diffusion of RSPS innovation, which showed high statistical significance for the differences between acceptors and rejecters in all four attributes. Acceptors appreciate the installation of RSPS for its *relative advantages*, while rejecters had a lower perception. However, these perceived advantages were the long-term benefits related to electricity costs. This includes reducing the effects of *global warming* as well as providing a source of *unlimited power* supply. Additionally, the study showed that the *installation cost* is the leading constraint that can drive participants to reject RSPS installation.

The generalization of these findings to other hydrocarbon-rich countries is guided by similarities in contexts, specifically a lower standard electricity prices and a similar per capita income. While other hydrocarbon-rich countries may share similar strategies considering transforming to other renewable energy sources, special consideration should be given to the lower standard electricity prices in Saudi Arabia compared to the solar-generated electricity price. More comparable costs in other countries may result in changing adopters' view for the *relative advantages* of RSPSs, which should be considered in innovation dissemination policies. In addition, as the adoption decision is individual at the family level, the perception of *installation cost*, as a limiting factor, may be correlated with the per capita income of a specific country. This research concludes that there is a requisite need to reconsider the current small-scale solar PV systems regulations and introduce information provision tools and incentive mechanisms to stimulate the adoption of RSPSs.

References

Abdullah, D., Zhou, D., Shah, T., Jebran, K., Ali, S., Ali, A., & Asad (2017). Acceptance and willingness to pay for solar home system: Survey evidence from northern area of Pakistan. *Energy Reports, 3*, 54–60. https://doi.org/10.1016/j.egyr.2017.03.002.

Akinboro, F., Adejumobi, L. A., & Makinde, V. (2012). Solar energy installation in Nigeria: observations, prospect, problems and solution. *Transnatl Journal Science Technology May Ed, 2*, 73–84.

Al Eqtisadiah [WWW Document]. (2017). 80 to 260% increase in electricity tariff for residential sector. Retrieved 7, Sep 2019, from http://www.aleqt.com/2017/12/13/article_1298286.html.

Al-Barrak, S. (2018). Personal communication with Shareef Al-Barrak, the Vice Director Technical Affairs of Electricity & Cogeneration Regulatory Authority ECRA, at the headquarters located in Riyadh, 2018.04.25.

Al-Ghamdi, S. A., & Alshaibani, K. A. (2018). The potential of solar energy in Saudi Arabia: The Residential Sector. *JEA*, 6. https://doi.org/10.15640/jea.v5n2a4.

Aliqtisadi (2016). Saudi Electricity consumes 2 million barrels of oil per day [WWW Document]. http://tiny.cc/bq6b8y (accessed 8.22.19).

Allen, S. R., Hammond, G. P., & McManus, M. C. (2008). *Applied Energy, 85*, 528–544. https://doi.org/10.1016/j.apenergy.2007.09.006.

Almasoud, A. H., & Gandayh, H. M. (2015). Future of solar energy in Saudi Arabia. *Journal of King Saud University—Engineering Sciences, 27,* 153–157. https://doi.org/10.1016/j.jksues.2014.03.007.

Al-Mutairi, K. A., Alfifi, A. H., Aljahni, S. B., & Albalawi, A. M. (2019). Climate changes knowledge and awareness among people in Tabuk region, Saudi Arabia. *ACTA SCIENTIFIC AGRICULTURE, 3,* 184–188.

Alrashed, F., & Asif, M. (2015). An exploratory of residents' views towards applying renewable energy systems in saudi dwellings. *Energy Procedia, 75,* 1341–1347. https://doi.org/10.1016/j.egypro.2015.07.207.

Alsubaie, A., & Jones, K. (2017). An overview of the current state of women's leadership in higher education in saudi arabia and a proposal for future research directions. *Administrative Sciences, 7,* 36. https://doi.org/10.3390/admsci7040036.

Al-Taher, M. (2018). Personal communication with Mohammed Y. Al-Taher, the Distribution Planning Department Manager of Saudi Electricity Company SEC, at the headquarters located in Riyadh, 2018.05.14.

AlYahya, S., & Irfan, M. (2016a). The techno-economic potential of Saudi Arabia's solar industry. *Renewable and Sustainable Energy Reviews, 55,* 697–702. https://doi.org/10.1016/j.rser.2015.11.017.

AlYahya, S., & Irfan, M. A. (2016b). Analysis from the new solar radiation Atlas for Saudi Arabia. *Solar Energy, 130,* 116–127. https://doi.org/10.1016/j.solener.2016.01.053.

Arthur D Little [WWW Document]. (2018). Saudi Arabia: World leader in solar power? Retrieved 21, Aug 2019, from https://www.adlittle.com/en/insights/viewpoints/saudi-arabia-world-leader-solar-power.

Asif, M. (2016). Urban scale application of solar PV to improve sustainability in the building and the energy sectors of KSA. *Sustainability, 8,* 1127. https://doi.org/10.3390/su8111127.

Barnett, H. G. (1953). *Innovation: The basis of cultural change.* New York, The United State of America: McGraw-Hill Book Company.

Beise, M. (2003). *Lead markets: Drivers of the global diffusion of innovations (Discussion Paper Series No. 141).* Research Institute for Economics & Business Administration, Kobe University.

Bridge, S. (2019). Saudi Arabia launches new $1.5bn phase of solar energy plan, Arabian Business news portal. Available online: [WWW Document]. ArabianBusiness.com. Retrieved 21, Aug 2019, from https://www.arabianbusiness.com/energy/412239-saudi-arabia-launches-new-15bn-phase-of-solar-energy-plan.

Byers, B. A. (1996). Understanding and Influencing Behaviors in Conservation and Natural Resources Management. The Nature Conservancy, and the World Resources Institute, Washington, DC, The United State of America.

Claudy, M., Driscoll, A. O.', & Duffy, A. (2010). Home Owners' Attitudes, Perceptions and Willingness to Pay for Microgeneration Technologies. https://doi.org/10.13140/2.1.3269.2483.

Clayton, V. (2012). KSA leads the world in mobile usage | | ITP.net [WWW Document]. Retrieved 29, Jan 2020, from https://www.itp.net/588729-ksa-leads-the-world-in-mobile-usage.

D'Agostino, R. B., & Stephens, M. A. (1986). *Goodness-of-fit techniques.* New York, The United State of America: Marcel Dekker, Inc.

Dehwah, A. H. A., Asif, M., & Rahman, M. T. (2018). Prospects of PV application in unregulated building rooftops in developing countries: A perspective from Saudi Arabia. *Energy and Buildings, 171,* 76–87. https://doi.org/10.1016/j.enbuild.2018.04.001.

DeVellis, R. F. (2016). *Scale development: Theory and applications.* New York, The United State of America: SAGE Publications Inc.

Electricity & Cogeneration Regulatory Authority ECRA (2017). Small-Scale Solar PV Systems Regulations.

Eltawil, M. A., & Zhao, Z. (2010). Grid-connected photovoltaic power systems: Technical and potential problems—A review. *Renewable and Sustainable Energy Reviews, 14,* 112–129. https://doi.org/10.1016/j.rser.2009.07.015.

Faiers, A., & Neame, C. (2006). Consumer attitudes towards domestic solar power systems. *Energy Policy, 34,* 1797–1806. https://doi.org/10.1016/j.enpol.2005.01.001.

Faiers, A., Neame, C., & Cook, M. (2007). The adoption of domestic solar-power systems: Do consumers assess product attributes in a stepwise process? *Energy Policy, 35,* 3418–3423. https://doi.org/10.1016/j.enpol.2006.10.029.

GAS (The total population of Saudi Arabia), 2018. General Authority of Statistics.

Geels, F. W. (2010). Ontologies, socio-technical transitions (to sustainability), and the multi-level perspective. *Research Policy, Special Section on Innovation and Sustainability Transitions, 39,* 495–510. https://doi.org/10.1016/j.respol.2010.01.022.

Hepbasli, A., & Alsuhaibani, Z. (2011). A key review on present status and future directions of solar energy studies and applications in Saudi Arabia. *Renewable and Sustainable Energy Reviews, 15,* 5021–5050. https://doi.org/10.1016/j.rser.2011.07.052.

Huijts, N. M. A., Molin, E. J. E., & Steg, L. (2012). Psychological factors influencing sustainable energy technology acceptance: A review-based comprehensive framework. *Renewable and Sustainable Energy Reviews, 16,* 525–531. https://doi.org/10.1016/j.rser.2011.08.018.

Jacobsson, S., & Lauber, V. (2006). The politics and policy of energy system transformation—explaining the German diffusion of renewable energy technology. *Energy Policy, Renewable Energy Policies in the European Union, 34,* 256–276. https://doi.org/10.1016/j.enpol.2004.08.029.

Joshi, A., Kale, S., Chandel, S., & Pal, D. (2015). Likert Scale: Explored and Explained. *BJAST, 7,* 396–403. https://doi.org/10.9734/BJAST/2015/14975.

Kamran, M., Mudassar, M., Ahmed, S. R., Mehmood, H. S., Joyia, F. M., Hussain, M., et al. (2019). Socio-Economic Acceptance for Stand-Alone Solar PV Systems: Survey Evidence from Southern Punjab, Pakistan. *International Journal of Renewable Energy Research (IJRER), 9,* 301–308.

Karakaya, E. (2015). *The case of solar photovoltaic systems in Germany and the disillusion of grid parity (doctoral degree).* Spain.

Karakaya, E., Hidalgo, A., & Nuur, C. (2014). Diffusion of eco-innovations: A review. *Renewable and Sustainable Energy Reviews, 33,* 392–399. https://doi.org/10.1016/j.rser.2014.01.083.

Khan, M. A., & Latif, N. (2010). Environmental friendly solar energy in Pakistan's scenario. *Renewable and Sustainable Energy Reviews, 14,* 2179–2181. https://doi.org/10.1016/j.rser.2010.03.016.

LEE, J. (1994). Odds ratio or relative risk for cross-Sectional Data? *International Journal of Epidemiology, 23,* 201–203. https://doi.org/10.1093/ije/23.1.201.

Lehman, A., O'Rourke, N., Hatcher, L., & Stepanski, E. (2005). *JMP for basic univariate and multivariate statistics: A step-by-step guide.* SAS Institute.

Leonard-Barton, D. (1980). The role of interpersonal communication networks in the diffusion of energy conserving practices and technologies. Presented at the The International Conference on Consumer Behavior and Energy Use, Banff, Alberta, Canada.

Likert, R. (1932). *A technique for the measurement of attitudes.* WOODIYORTE, Eorron, New York, The United State of America: ARCHIVES OF PSYCHOLOGY. R. S.

MobileGate. Study: Saudi Arabia ranks first in the list of countries most used for smartphones in the world [WWW Document], 2017. Retrieved 22, Aug 2019, from https://bit.ly/2MocNeG.

Mosly, I., & Makki, A. (2018). Current status and willingness to adopt renewable energy technologies in Saudi Arabia. *Sustainability, 10,* 4269. https://doi.org/10.3390/su10114269.

Muntasser, M. A., Bara, M. F., Quadri, H. A., EL-Tarabelsi, R., & La-azebi, I. F. (2000). Photovoltaic marketing in developing countries. *Applied Energy, 65,* 67–72. https://doi.org/10.1016/S0306-2619(99)00094-X.

Neame, C. (2009). *Understanding the adoption of solar power technologies in the UK domestic sector (doctoral degree).* Cranfield, The United Kingdom: Cranfield University.

Oxford Dictionaries [WWW Document]. (2012). Definition of acceptance in English. https://www.lexico.com/definition/acceptance.

Poonnavich, S., & Weerin, W. (2015). Critical Factors of Social Acceptance for Solar PV Rooftop in Thailand. Presented at the 2015 4th International Conference on Informatics, Environment, Energy and Applications, Pattaya, Thailand. https://doi.org/10.7763/IPCBEE.

Qureshi, T. M., Ullah, K., & Arentsen, M. J. (2017). Factors responsible for solar PV adoption at household level: A case of Lahore, Pakistan. *Renewable and Sustainable Energy Reviews, 78,* 754–763. https://doi.org/10.1016/j.rser.2017.04.020.

Ramli, M. A. M., Hiendro, A., & Al-Turki, Y. A. (2016). Techno-economic energy analysis of wind/solar hybrid system: Case study for western coastal area of Saudi Arabia. *Renewable Energy, 91,* 374–385. https://doi.org/10.1016/j.renene.2016.01.071.

Ratikainen, K. W. (2017). *Transitioning to renewable energy in Saudi Arabia, A multi-level perspective analysis of the Saudi renewable energy policies*. University of Oslo, Norway.

Renewable Energy Project Development Office. Ministry of Energy, Industry and Mineral Resources [WWW Document]. (2017). Retrieved 21, Aug 2019, from https://www.powersaudiarabia.com.sa/web/index.html.

Robertson, T. S. (1967). The process of innovation and the diffusion of innovation. *Journal Marking, 31,* 14–19. https://doi.org/10.1177/002224296703100104.

Rogers, E. M. (1983). *Diffusion of innovations*, 3rd Edition, 3rd ed. Free Press ; Collier Macmillan, New York : London.

Rogers, E. M. (1995). *Diffusion of innovations* (4th ed.). New York: Free Press.

Rogers, E. M. (2003). *Diffusion of Innovations* (5th ed.). New York: Free Press.

Rogers, E., & Adhikarya, R. (1979). Diffusion of innovations: An Up-to-date review and commentary. In *Communication Yearbook 3. The International Communication Association, New Brunswick, New Jersey, The United State of America.*

Salam, M. A., & Khan, S. A. (2018). Transition towards sustainable energy production—A review of the progress for solar energy in Saudi Arabia. *Energy Exploration & Exploitation, 36,* 3–27. https://doi.org/10.1177/0144598717737442.

Saudi Vision (2016). Saudi Vision 2030, Media Center. [WWW Document]. https://vision2030.gov.sa/en/media-center (accessed 7.23.19).

Schumpeter, J. A. (1939). *Business cycles McGraw Hill book company*. New York, The United State of America: McGraw Hill Book Company.

SEC [WWW Document]. (2015). Saudi Electricity Company strategy. Retrieved 28, July 2019, from https://www.se.com.sa/ar-sa/invshareholder/Pages/strategy.aspx.

Seznec, J.-F. (2018). *Renewable energy in the middle East*. Global Energy Center, Washington, DC, The United State of America: Atlantic Council.

Solangi, K. H., Saidur, R., Luhur, M. R., Aman, M. M., Badarudin, A., Kazi, S. N., et al. (2015). Social acceptance of solar energy in Malaysia: users' perspective. *Clean Techn Environ Policy, 17,* 1975–1986. https://doi.org/10.1007/s10098-015-0920-2.

Solvak, M., Unt, T., Rozgonjuk, D., Võrk, A., Veskimäe, M., & Vassil, K. (2019). E-governance diffusion: Population level e-service adoption rates and usage patterns. *Telematics and Informatics, 36,* 39–54. https://doi.org/10.1016/j.tele.2018.11.005.

Tareq, F. (2017). Acceptance and willingness to pay for pv solar electricity: survey evidence from southern part of Bangladesh. *Presented at the International Conference on Mechanical Engineering and Renewable Energy, Chittagong, Bangladesh*, p. 7.

Tidd, J. (2010). From models to the management of diffusion. In *Gaining Momentum, Series on Technology Management. IMPERIAL COLLEGE PRESS*, pp. 3–45. https://doi.org/10.1142/9781848163553_0001.

Tlili, I. (2015). Renewable energy in Saudi Arabia: current status and future potentials. *Environment, Development and Sustainability, 17,* 859–886. https://doi.org/10.1007/s10668-014-9579-9.

Vasseu, V., & Kemp, R. (2015). The adoption of PV in the Netherlands: A statistical analysis of adoption factors. *Renewable and Sustainable Energy Reviews, 41,* 483–494. https://doi.org/10.1016/j.rser.2014.08.020.

Wamukonya, N. (2003). Power sector reform in developing countries: Mismatched agendas. *Energy Policy, 31,* 1273–1289. https://doi.org/10.1016/S0301-4215(02)00187-8.

Yin, R. K. (2009). *Case study research: design and methods*. SAGE Publications, Inc.

Chapter 5
Multi-level Perspective of Technological Transitions: Analysis of the Residential Solar Photovoltaic Systems Policies in Hydrocarbon-Rich Saudi Arabia

In this chapter, the *multi-level perspective* (*MLP*) was used as an analytical tool to capture the socio-technical transition of RSPSs by integrating different explanatory models in various dimensions. A questionnaire was developed based on the *MLP* three integrated layers of socio-technical change: Niche (micro-level), Regime (meso-level), and Landscape (macro-level), to elucidate the challenges that the RSPSs transition is facing in the rentier Saudi state. A total of 32 face-to-face semi-structured interviews were conducted with stakeholders of the renewable energy sector in Saudi Arabia. The underlying goal of these interviews is to gather first-hand information and informed opinions on the current transition in the solar energy landscape in Saudi Arabia, preparing the ground to integrate *transition management* (*TM*) policy design model for the case of the RSPSs.

5.1 Research Background

Over the past several decades, enormous revenues derived from hydrocarbon exports have been a fundamental element in the development and prosperity of Saudi Arabia. They allowed the country to expand its influence beyond energy markets to play a political and economic pivotal role, both at regional and international levels. However, as the world's largest oil exporter country, the unsustainable domestic growth, which is fundamentally built on fossil fuel exports, increases the threats to not only the domestic economy but also the world economy, OEC (2019).

This urged the Saudi government to take significant steps to remedy the economy's distortions by diversifying the sources of its national income and redesigning a new energy-mix that incorporates shares of renewable energy. Among various measures have been taken recently to ease the pressure on the government's resources is to abolish electricity subsidy and enable individual households to generate a share of the electricity they need through small-scale distributed RSPSs. Besides the significantly

K. A. Alrashoud, *Managing Technology Transition in Saudi Arabia*,
https://doi.org/10.1007/978-981-15-5803-0_5

small number of adopters, these kinds of approaches are critical, especially for a country such as Saudi Arabia. In other words, substantial revenues from hydrocarbons exports have facilitated what has been referred to as the "ruling bargain". Low-priced utility services represent one of the most significant features of this social contract, and changes in this domain influence not only the country's socio-technical system but also the socio-political system.

The political structure is critical for any technological transition, and due to the complexity of the interlinking associations between the levels and processes of innovation diffusion, an in-depth descriptive investigation is needed to draw a comprehensive picture of the RSPSs transition in Saudi Arabia. This study aims to analyze key material dimensions of solar energy policies in Saudi Arabia, with a particular focus on the RSPSs implementation. Since sustainability transition-related studies that were drawn upon theoretical concepts have only rarely been conducted in this context, the present study will be a significant endeavor that redounds to the benefit of the policymakers, the local communities, and solar project developers. It will also serve as a future reference for research on the subject of RSPSs dissemination in hydrocarbon-rich countries.

5.2 Socio-Technical Transitions to Sustainability

Sustainability transitions are known for being long-term, multi-dimensional transformation processes that take place within socio-technical systems to shift to more sustainable conditions (Jochen et al. 2012). Various scholars have indicated that shifting towards sustainability requires changes not only in technology, but also in societal structures, behavior patterns, and cultural practices (Jochen et al. 2012; John et al. 2010). These changes need to be guided and managed by interdisciplinary methods of research that consider the interactions between institutional, technological, cultural, and environmental subsystems (Binder et al. 2013, 2017; Geels 2011; Young et al. 2006).

As the layout of the world's energy systems has changed substantially over time, national systems have become more integrated (Häfele and Sassin 1977). Energy transition, from a fossil fuel-based to a more green energy supply system, might be considered as the most complex and long-term transaction in any society (John et al. 2010). Although numerous researches have been conducted on renewable energy transitions over the past two decades, these researchers have been primarily conducted in the context of developed western countries (Normann 2015; Osunmuyiwa et al. 2018; Rolffs et al. 2015; Ulsrud et al. 2011). Focusing on the analytical aspect, many of these studies addressed basic questions of how a systemic change is examined and whether a social system can be driven into particular trajectories. These notions have been studied broadly by Albert (1995), Kiel and Elliott (1997), Wimmer and Kössler (2005). While the editions of the first two innovatively examine the inter-relationships between the chaos theory and the social transition, the latter has documented theories of a cross-disciplinary dialogue of models, metaphors, and methodologies for understanding change.

On a slightly different level, Schäpke (2018) addresses the fundamental tension between transitions as an open-ended process of change and the normative directions of this change towards sustainability. Schäpke chooses the salient *TM* policy design model and *MLP* to serve as a study framework and application context. By conducting a conceptual-theoretical study that focuses on three themes of sustainability transitions (individual agency, normativity, and transdisciplinary collaboration), Schäpke proposed several principles to understand sustainability transition management better. However, since Schäpke does not include a regional focus, his work lacks a dimension that is essential to any technology transition.

While energy transition studies conducted in the context of the Arab region are minimal, one of the most substantial academic work done on renewable energy transition in the Middle Eastern context was by Kumetat (2012). Using both the *MLP* and *TM* approaches in targeting the elements of socio-technical governance theory, energy policy analysis, and Middle Eastern studies, Kumetat further developed an analytical framework specific to hydrocarbon-rich countries, by which he conducted a comparative case study on the Democratic Republic of Algeria and the United Arab Emirates. Through the results of interviewing elites in the respective countries, the in-depth interdisciplinary analysis, conducted by Kumetat, reveals insights into renewable energy policies in these hydrocarbon-rich Arab states.

Researchers of sustainability transitions tended to address renewable energy systems and overlook the distinctive nature of RSPSs and its implication for other regime actors. To our knowledge, there is not a comprehensive Saudi Arabia-specific analysis of the use in, and consequently appropriateness to RSPSs of the *MLP*. To address this gap, the present study provides a critical analysis by casting light on the RSPSs policies based on the three levels of the *MLP* (landscape level, regime level, and niche level).

5.3 Conceptual Framework: Multi-level Perspective and Policy Design Modeling of Transition Management

While several theories analyze societal transitions, *MLP* emerges as one of the prominent comprehensive analytical tools used to understand the dynamics of such transitions. *MLP* was initially developed by Rip et al. (1998) then further refined by Geels (2005, Geels and Schot (2007, John et al. (2010). This analytical tool not only helps to understand the transition in terms of technological innovation but also analyzes the societal factors in transition trajectories. As illustrated in Fig. 5.1, *MLP* identifies three levels in an innovation transition process; Niche (micro-level), Regime (meso-level), and Landscape (macro-level) (Geels 2002).

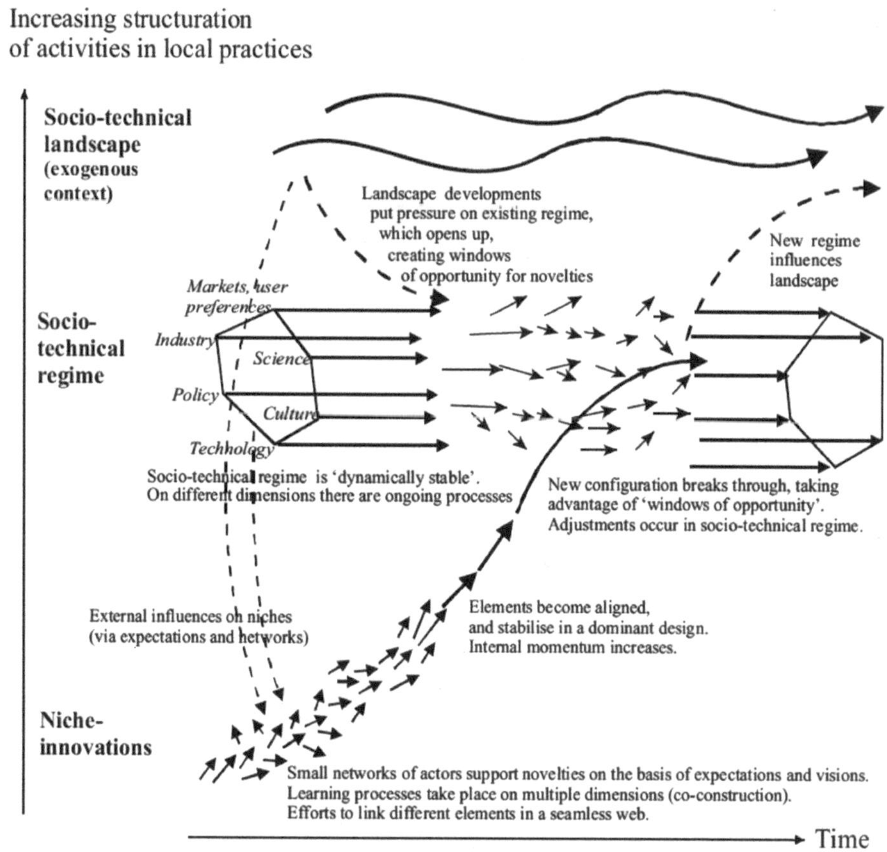

Fig. 5.1 The multi-level perspective on transitions. *Source* Geels (2002)

Niche level is where space and time are provided to allow for radical innovation to happen. This level acts as incubation rooms for innovation to be developed by research and learning through experience. On the other hand, the Regime level is the core of *MLP* representing the current structures of a system (e.g., energy systems), including the institutions, associations, and technologies that are self-reinforcing. It encompasses social groups and actors, as well as formal and cognitive rules. The landscape level is, unlike the regime level, known as the exogenous broader for contextual developments that embody macro-politics, macro-economic and other external pressure factors such as wars, natural disasters, oil price shocks, and political upheavals.

It is noteworthy that Geels proposes five inner socio-technical (ST) regimes within the meso level, where institutions/rules interlink with and between other regimes (Geels 2004). These ST regimes are technological regime, user and market regime, socio-cultural regime, policy regime, and science regime (Fig. 5.2).

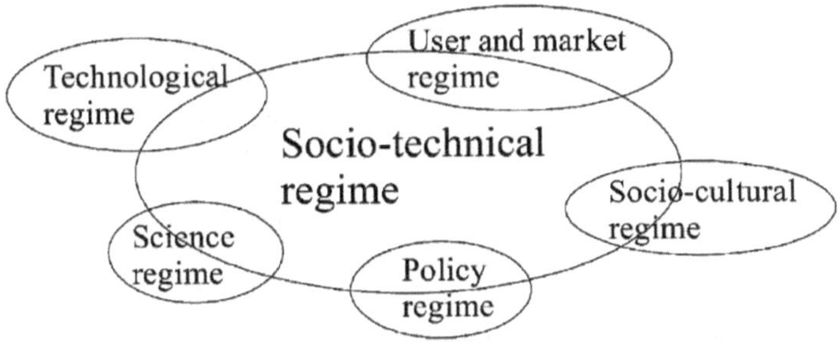

Fig. 5.2 The five socio-technical regimes. *Source* Geels (2004)

Despite its strengths, the *MLP* received criticisms against its validity, which need to be considered (e.g., flat ontology, lack of agency, operationalization of regimes, bias towards bottom-up change models, and epistemology and explanatory style). One critique targets the 'regime' conceptualization for its failure to account for issues of agency and the politics of transitions (Grin et al. 2011; Meadowcroft 2011). *MLP* recognizes the interests and the role power of the regime's actors. However, due to its theoretical based structure, *MLP* has so far paid less focus on the scales and roles of collective actors (Geels 2014). In response to these critiques, several studies have been conducted, expanding a more comprehensive range of theoretical and empirical explanations in understanding the interlinking role of actors, power, agency, and politics within the regime. Extended responses to the criticisms against *MLP* can be found in Geels' work who has collected and addressed the main seven critiques against the validity of the *MLP* (Geels 2011). As augured by Geels and Schot (2007), the *MLP* functional role serves as a tool providing "narrative explanations" of a larger picture of patterns that exist within and between socio-technical levels, rather than identifying empirical data or specific variables. Suggesting very general statements to describe the development of a certain level (e.g. climate change boost renewable energy policy) is considered as one limitation of the *MLP*. Acknowledging this fact, however, such macro-drivers, including region-specific drivers, need to be considered, which this study does on a qualitative approach across stakeholder interview data.

Although it is elusive to develop a framework that embeds all elements to capture a socio-technical transition, it is essential to consider the context of the case being studied. The analytical dimensions framework of renewable energy in hydrocarbon-rich Arab countries developed by Kumetat (2012) provides the basis of this study, as it offers an overreaching structure for the policy analysis of RSPSs in Saudi Arabia. Table 5.1 presents the analytical dimensions of the RSPSs policies in Saudi Arabia. The descriptive value and comprehensiveness of this framework will be tested by applying it to the studied case.

While the *MLP* is a powerful explanatory tool laying a strong focus on system structures and the role of actors, it does not describe processes of transition. Instead, it highlights its character as a hermeneutic tool of the multifariousness and complexity of socio-technical change (Geels 2002), and is, therefore, in its refined form, an

Table 5.1 Analytical dimensions of the RSPSs policies in hydrocarbon-rich Saudi Arabia structured

Scale	Definition	Analytical focus
Landscape (macro-level)	Long-term macro-drivers that cannot effortlessly be altered by regime-level stakeholders	• Political system • Key regional and transregional energy governance bodies • Climate change • Long-term national economic diversifications • Country branding
Regime (meso-level)	Entrenched actors and dominant regulatory, financial, and political systems	• *Technological regime*: the readiness of the national power grid, including power generation, transmission, and distribution; nuclear energy option • *User and market regime*: structure of national energy markets, legal and institutional frameworks • *Socio-cultural regime*: consumption patterns, public environmental awareness, and a potential role of religion ethics • *Policy regime*: national targets of the RSPSs and actual performance, government's energy subsidies • *Science regime*: technology transfer, industry structures, R&D strategies
Niche (micro-level)	Protected spaces with a potential for radical change in the established regimes	• Technological developments and local content enhancement

Adopted from Kumetat (2012), modified by the author

appropriate tool to analyze the RSPSs policies in hydrocarbon-rich Saudi Arabia. Due to the descriptive and exploratory style of this study, which attempts to understand how should the energy policies of RSPSs and the transition towards it in Saudi Arabia be understood? A total of 19 research subsidiary questions (sub-questions) were developed to fit into the Saudi context while maintaining the scale of the *MLP* model. A full compilation of these sub-questions which reflect each key analytical element can be given as follow:

Landscape (macro-level)

- Political system

SQ1. To what extent do the political system and the non-existence of independent Civil Society Organizations allow for system innovations in the solar energy sector, particularly bottom-up innovations?

SQ2. How likely is the Saudi Vision 2030 goals for solar energy to be altered by changes in the political system?

SQ3. To what extent are the formal decision-making processes the actual decision-making structures?

SQ4. Which role do transregional energy governance bodies such as OPEC, OAPEC, or IRENA play in the promotion or stalling of national solar energy policy in Saudi Arabia?

- Climate Change

SQ5. Is the climate change agenda actively considered in the political decision-making on renewable energy developments in Saudi Arabia?

- Long-term national economic diversification

SQ6. To what extent are oil prices related to the deployment of solar energy in Saudi Arabia?

- Country branding

SQ7. Is country branding seen as a motive in the spread of solar energy?

Regime (meso-level)

- Technological regime

SQ8. To what extent power generation, transmission, and distribution infrastructure in Saudi Arabia are sufficient for a successful deployment of residential solar photovoltaic systems?

SQ9. Is nuclear energy seen as a competitor of solar power production?

- User and market regime

SQ10. Is there any plan for residential solar photovoltaic systems promotion in place?

SQ11. Do you believe the transfer of European renewable energy policy design models is a viable method to spread residential solar electricity in Saudi Arabia, or would indigenous policy models work better?

SQ12. What is the most significant factor that may act as a barrier to the deployment of residential solar systems?

- Socio-cultural regime

SQ13. How widespread is the knowledge of residential solar systems among individuals in Saudi Arabia?

SQ14. Does religion play a role in promoting the use of solar energy technology?

- Policy regime

SQ15. What is the residential solar systems' share of the total solar energy capacity to be installed by 2030?

SQ16. How seriously do the government bodies consider the social aspect when removing energy subsidies?

- Science regime

SQ17. Is there an attempt to create a domestic knowledge hub for renewable energy technologies, and is there an industry structure that supports the spread of technologies into mainstream society?

Niche (micro-level)

- Technological developments

SQ18. What is your opinion on technology choices and the potential of the solar energy industry in Saudi Arabia?

SQ19. How prepared is the country to develop a robust Local Content in the solar power sector?

While the MLP analytical framework for the Hydrocarbon-rich Arab countries is considered an inclusive structure for policy analysis in this context, the author of this book conceivably modified the questions of this theoretical framework to appropriately suit the temporal and spatial context of the case investigated in this study.

Table 5.2 presented below highlights the key modifications made to the theoretical framework.

Despite the fact the *MLP* is a well-established hermeneutic analytical tool of energy transitions, its contribution to the formulation of policies and regulations remains inadequate. Therefore, the *TM* approach, which is closely interlinked with the *MLP* framework, will be used as a model of policy design to test its applicability in the case of RSPSs in Saudi Arabia.

TM is a governance approach that seeks to facilitate a transition to sustainability through a participatory process of visioning, experimenting, and learning (Foxton et al. 2008; Rotmans et al. 2001). Opinions have varied on how the beginning of *TM* has evolved. One theorist considers the sociological facet of transition as deeply entrenched within-population dynamics (Rotmans et al. 2001). Other schools attribute the development of *TM* to the link between the social and technical factors of the co-evolution and the systems theory (Elizabeth and Gordon 2007). Most agree that shifts from power-centralized governments to more liberal market-driven governments have allowed new patters of bottom-up governance forms to rise and play more significant roles within regime systems (Loorbach 2010). *TM* was used broadly by the Dutch government which sought to manage the radical transition of the domestic

Table 5.2 Key modification to the analytical framework developed by Kumetat for the Hydrocarbon-rich Arab countries

Scale	Analytical focus	Kumetat (2012)		The author		Geels (2004)'s definition
		Respective question(s)	Explanation	Respective question(s)	Justification	
Landscape level	Political system and long-term effects of renewables	SQ3	The raised question is too generic, most likely to be applicable to both of the two cases studied, UAE and Algeria	SQ2	More specificity is added, linking the country's national vision for solar energy with potential changes in the political system	–
	Long-term national economic diversification	SQ6	Inquiring the connectivity between renewable energy-related efforts and the national economic visions in the respective countries studied	SQ6	As a national vision that encompasses solar energy has already been formed, a question about the influence of oil prices on the deployment of solar energy was raised	–
Regime level	Technological regime	SQ8	Inquiring about the extent to which renewable energies are employed to reduce pressure on conventional power plans in each respective county (Federal structure of a political system in the UAE and democratic nature of a system in Algeria)	SQ8	Despite the high similarity of the economic features between Saudi Arabia, UAE, and Algeria, the political system in Saudi Arabia is slightly different. As a monarchy with no autonomous cities, the installed capacity of renewable energy is identified in the literature. Instead, a sub-question was raised about the national power grid's readiness for the penetration of renewable energy	Geels identifies the technological regime as: "the rule-set or grammar embedded in a complex of engineering practices, production process technologies, product characteristics, skills and procedures, ways of handling relevant artefacts and persons, ways of defining problems; all of them embedded in institutions and infrastructures"

(continued)

Table 5.2 (continued)

Scale	Analytical focus	Kumetat (2012)		The author		Geels (2004)'s definition
		Respective question(s)	Explanation	Respective question(s)	Justification	
	User and market regime	–	–	SQ12	A straightforward question that falls into the domain of the regime level in which the other sub-systems interlink with each other	–
	Science regime	SQ17	On the contrary to Geels's proposal, Kumetat classified the science regime within the meso-level of analysis. Kumetat argues that questions that investigate whether there are efforts to create a domestic knowledge hub or industry structure that supports the deployment of renewable energy technologies fits more into the micro-level	SQ17	Given the nature of the kingdom's energy systems and its political and economic structures, such initiatives are unlikely to be developed from the micro-level. This fact is also, to some extent, supported by the findings of Kumetat's study, where the bottom-up innovations structures in the UAE and Algeria found to be very limited. Above all, classifying such a dimension into the micro-level fundamentally contrasts with the essence of the science regime as proposed by Geels (Fig. 5.2)	According to Geels, some examples of the *science regime*'s formal rules are "Formal research programmes (in research groups, governments), professional boundaries, rules for government subsidies"
Niche level	–	–	–	SQ19	Technology localization is a fundamental element of any development of technologies. A research sub-question was raised to investigate the trajectories of this key matter, as Saudi Arabia has an abundant stock of pure silica sand which is considered a key component of the manufacturing of PV panels	–

energy systems in the 2000s. In close cooperation with the researchers (Geels 2004; Kemp et al. 2007b; Rotmans et al. 2001; Verbong and Geels 2007), the *TM* was implemented as a part of the fourth National Environmental Policy Plan in the Netherlands (Kemp 2010; Rotmans et al. 2001).

TM is a cyclical process that integrates different substances and clusters of activities with a great deal of flexibility. It starts with the phase of problem identification and vision development, further development of transition agendas, implantation transition experiments, evaluation and adaptation and iteration of transient processes (Fig. 5.3).

Following key transitions literature of Kemp et al. (2007a), Kumetat (2012), Loorbach (2007), a TM process of the case of RSPSs diffusion in Saudi Arabia would entail the following steps:

(1) Problem structuring in the transition arena and formulating a vision
(2) Devising a concrete transition pathway and agenda
(3) Experimental phase
(4) Monitor and evaluate the results

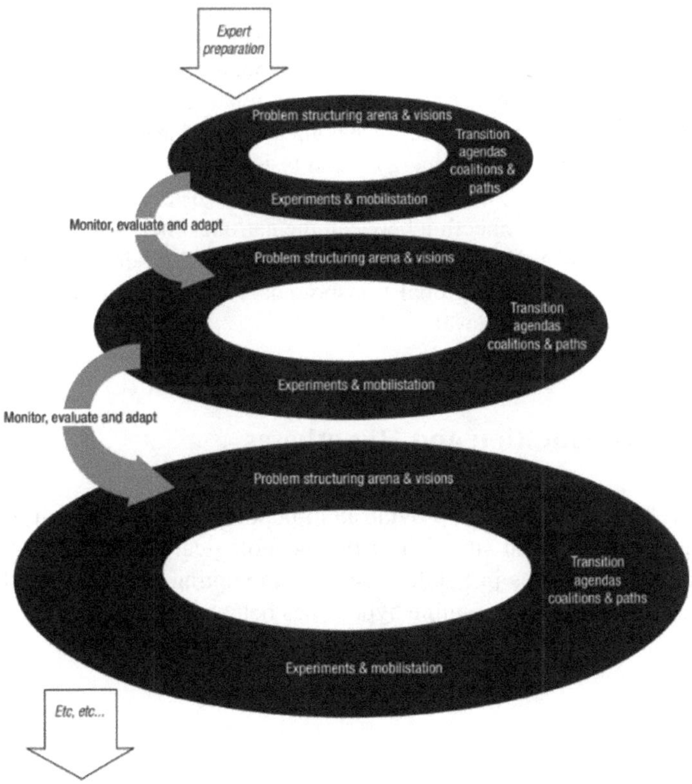

Fig. 5.3 Cycles of transition management. *Source* Loorbach (2007)

(5) Adaptation
(6) Reiterate from step (1).

Theorists have risen several criticisms against the *TM* model that need to be considered.[1] For instance, the idea of formulating a guiding vision by which government policies are modeled has received critiques by Berkhout et al. (2004), arguing that the results of many unsuccessful socialist economic plans show that master plans have rarely guided innovations. Considering the complex nature of innovation, one could argue the difficulty of introducing one comprehensive central plan. While (Rotmans and Loorbach 2008) slightly counter this argument, they emphasize that a vision of the *TM* sense should not be formulated as a step-by-step instruction manual that guides the users to achieve ultimate objectives. But instead, the *TM* vision should be much broader to inspire more technical plans which ultimately assists in painting an overall picture of the socio-technical transition.

Others argue that *TM* reflects open, liberal governments that maintain a high level of flexibility and adaptive capacity (Rotmans and Loorbach 2008). On the other hand, Kern claims that the *TM* model does not significantly reflect national politics and power relations, which makes it applicable in other contexts (Kern 2010).

An argument over the national level and the appropriate system boundaries for policy process and analysis have been raised by Meadowcroft (2005). While it is true that social innovations and technological advancements tend to be marked-based, led by global trends, and often not subjected to the national level, in this study, however, energy systems are considered as reference points. This is due to the fact that subsidies for conventional energy sources, energy markets, electricity demand and vital actors of the field are all, indeed, organized within a national level. This certainly does not mean neglecting the connection between the national energy systems and global energy developments. Acknowledging this fact, the *MLP* model will be used as it, to a certain extent, integrates global transboundary phenomena into the scope of analysis within its landscape level.

5.4 Research Question and Hypotheses

By obtaining first-hand data that provide an in-depth descriptive capacity, this study aims to analyze the current situation of the technological transition of RSPSs and factors shaping its policies in Saudi Arabia by attempting to answer the following main research question and examine hypotheses below.

[1]Extensive discussion on critiques towards transition managements is provided by Rotmans and Loorbach (2008).

5.4.1 *Research Question*

While the question shown below which is referred to as "RQ4", is the main research question of this chapter, the 19 sub-questions presented in Sect. 5.3 tailored the interview questions that were asked in the interviews.

RQ4. How should the energy policies of the residential solar PV systems and the transition towards it in Saudi Arabia be understood?

5.4.2 *Research Hypotheses*

(H5) There are no identified goals of RSPS planned capacity or how it should be implemented in alignment with the national renewable energy plans.

(H6) The participation of the independent Civil Society Organizations in the formulation of the RSPSs policies is significantly limited, with a neglect of the soft-power elements, such as religion and country branding as potential promotion tools.

5.5 Methodology and Research Design

Given the explorative character of this study, the case study design was selected as a suitable research strategy to achieve the objectives of this study. Case study design helps to identify patterns and trends within particular phenomena of development. It is preferred to examine contemporary events when relevant behaviors cannot be manipulated. This design can also have various epistemological aims and exceed the descriptive level and go beyond to contribute to theory formulation (Yin 2009). This research does not consider itself a study on energy pricing or a technical-related aspect of solar energy generation, but rather on the stakeholders' perception and judgment about governance concepts that are of empirical concern, which cannot be identified by a quantitative approach.

Additionally, some aspects related to the energy policy formation in the Arab countries in general, and the GCC countries, in particular, are considered sensitives topics by many regime stakeholders. Thus, a private research-interview with stakeholders has more chances of success in extracting information than collecting information from less personal methods or written texts. Therefore, in-depth semi-structured personal interviews were conducted employing the *MLP* to draw a comprehensive picture of the RSPSs technology transition in Saudi Arabia. A total of 32 interviews[2] were conducted with stakeholders in the renewable energy sector in Saudi Arabia

[2]Some interview events consisted of two or more interviews due to the interviewee's availability. These events have been counted as one interview. Similarly, in one case, two interviewees have

Table 5.3 Interviewees' backgrounds and approach methods

Category	Variable	No. of interviews
Nationality	Saudi	28
	Non-Saudi	4
Affiliation	Public stakeholder	5
	Private stakeholder	13
	R&D stakeholder	14
Method	Face-to-face	26
	Telephone	6

between the period of Mach 2018 to May 2018. Amongst the interviewees were 28 Saudi citizens stakeholders and 4 Non-Saudi stakeholders (Table 5.3).

The interviewees' population was divided into three categories: Public stakeholders, Private stakeholders, and Research and Development (R&D) stakeholders. The selection criterion of the interviewee requires an interviewee to have either significant influence over the renewable energy policy or directly be affected by these policies.

The author exerted many efforts to meet with the highest-ranked and well-informed members as possible. The access to those interviewees was achieved through the authors' personal connections with a number of energy specialists in Saudi Arabia. Additionally, more specialists have been approached and connections with them have been developed for the interviews through the professional networking platform, LinkedIn. Snowballing technique was also applied to tap into the respective business communities and personal interviews have been facilitated.

Although the face-to-face approach was used to obtain information from the interviewees, six interviews were conducted by telephone calls due to the stakeholder's tight schedules or access constraints. Three out of the six telephone calls were made after meeting personally with the perspective stakeholders yet were counted as telephone interviews as all questions were asked and answered through telephone calls. …"[…] telephone interview has a specific advantage that a second call can be easily arranged if the interview has to be rescheduled, or the interview can be more conveniently timed for participants in different time zones" (Musselwhite et al. 2007). Although the observation of gestures and body-language is impossible through telephone interviews, the tone of voice could serve as an alternative valuation indicator of the soundness of the respondents' answers (Frey and Oishi 1995).

At the beginning of the interview, the briefing phase kick starts, where the context of the interview is described and aims are highlighted. Next, the warm-up phase follows, where the author approaches the topic and explains the interview framework which by the questions were asked. The author would handle the interviewees a hard copy of a document that contains information about the research background and the interview guideline. The author would also have a similar document to take

occasionally participated in the same interview. They have been counted as a one stakeholder since they belong to the same affiliation and had matched views.

notes beside the voice-recording. A verbal explanation was given to those who were interviewed thorough telephone calls. The third phase is the main phase where the questions are asked, and primary data is acquired. Next is the debriefing phase where the author made an effort to probe, clarify and reinforce the respondents. Last is the closing phase which by the interview is wrapped. The outcomes of both face-to-face and telephone semi-structured interviews were transcribed, coded, analyzed and categorized manually by identifying similarities and differences within the content of the transcriptions.

All interviewees completed answering the subsidiary research questions that were developed to cover all three analytical and heuristic levels posited by the *MLP* framework; Landscape (macro-level), Regime (meso-level), and Niche (micro-level). Table 5.4 shows the three *multi-level perspective* and its analytical dimensions that are covered separate subsections.

It should be pointed out that some of the interview questions deal with topics, which is a country like Saudi Arabia could be seen sensitive. For instance, some respondents may try to avoid or give unclear answers to questions that politically sensitive. This is an area where observance of ethical practices is especially relevant. As indicated by Renzetti and Lee (1993), when discussing sensitives topics, investigators must assure the respondents' anonymity (Renzetti and Lee 1993). In this work, all respondents had the freedom to select a place to conduct the interview and disclose or conceal their names and affiliations. Also, some of the interviewees exhibited their preference for not voice-recording the interviewer, but rather limit it to only note-taking. These techniques were applied accordingly, and consequently, helped to build trust with the respondents and increased their probability of having

Table 5.4 Analytical dimensions structured by subsections

Scale	Analytical focus	Sections
Landscape (macro-level)	• Political system • Key regional and transregional energy governance bodies • Climate change • Long-term national economic diversifications • Country branding	• 5.6.1.1 • 5.6.1.2 • 5.6.1.3 • 5.6.1.4 • 5.6.1.5
Regime (meso-level)	• *Technological regime*: the readiness of the national power grid, including power generation, transmission, and distribution; nuclear energy option • *User and market regime*: structure of national energy markets, legal and institutional frameworks • *Socio-cultural regime*: consumption patterns, public environmental awareness, and a potential role of religion ethics • *Policy regime*: national targets of the RSPSs and actual performance, government's energy subsidies • *Science regime*: technology transfer, industry structures, R&D strategies	• 5.6.2.1 • 5.6.2.2 • 5.6.2.3 • 5.6.2.4 • 5.6.2.5
Niche (micro-level)	• Technological developments and local content enhancement	• 5.6.3.1

Table 5.5 Interviewees category definition

Category	Definition
Public stakeholders	Renewable energy policymakers, state-owned grid operators, power companies, members of administrations or regulatory agencies
Private stakeholders	CEOs, owners or representatives of renewable energy companies, private investors
R&D stakeholders	Independent think tanks, in-house researchers and consultants, academics, scientific research governmental entities with separate legal personality

them to express their opinions more openly. An additional limitation to the face-to-face interviews is the difficulty of interviewing high-ranking government officials and corporates CEOs due to their busy schedules.

In some cases, researchers and investigators may get referred to another member who occupies a less senior position. Although this issue cannot fully be solved, the author who conducted this survey in the same country made an effort to overcome this challenge. This includes taking advantage of the existing personal contacts, developing new contacts from the early stages. As being Saudi national, the author had the liberty to move freely within Saudi Arabia and stayed as long as possible to set meetings with the targeted interviewees.

The total duration of the interviews' voice-recording was approximately 1090[3] min, averaging in one-hour interview per each stakeholder, and 135 pages of transcriptions. The underlying goal of these interviews is to gather informed opinions on the current transition in the solar energy landscape in Saudi Arabia.

This study limited itself to the application of RSPSs. Thus, all other renewable energy sources (such as wind, biomass, and geothermal energy, etc.) were excluded as they presently have little or no commercial significance or potentials in Saudi Arabia.

5.5.1 Interviewees Categorization

To improve the structure of the interviews and to enhance its outcomes, the population of the interviewees was divided into three defining categories: public stakeholders, private stakeholders, and R&D stakeholders. Each category has a specific definition, as shown in (Table 5.5), where all the interviews were coded and categorized based on the interviewees' affiliation.

[3]Five interviewees exhibited their preference for not to be voice-recorded. Thus, the note-taking technique, instead, was used.

5.5.1.1 Public Stakeholders

Public stakeholders' category encompasses those who are considered to be policymakers of renewable energy with an official public mandate or those who are directly involved in regulations or policy processes. This category includes members of the respective administrations, regulatory agencies, state-owned grid operators, and state-owned power companies. Representative elites of entities including the Ministry of Energy,[4] King Abdullah City for Atomic and Renewable Energy, Electricity & Cogeneration Regulatory Authority, and Saudi Electricity Company were interviewed in this category.

5.5.1.2 Private Stakeholders

Private stakeholders' group is a collective category for renewable energy companies' owners, CEOs, or senior-level representatives. This comprises all the renewable energy-related private companies involved in manufacturing, power production, or renewable energy spare parts retail. Private investors in all the mentioned-above areas, as well as private energy consultants, are also included in this category. High-ranking representatives from corporates including but not limited to Aramco, ACWA Power, Desert Technologies, NADEC, assigned to this category, were interviewed.

5.5.1.3 R&D Stakeholders

The common denominator of the R&D stakeholders' category is that all members are affiliated with profit/non-profit, independent/governmental think tanks or research centers or universities. All interviewees in this category are directly involved in researches concerning renewable energy, with a special focus on solar energy applications and implementation.

Due to the dynamic changes in the energy landscape in Saudi Arabia is witnessing, it has to be highlighted that the three categories and their role may overlap at certain levels. Although this matter cannot be fully resolved, entities and their actors have been assigned to the group where they fit most.

Although the author has successfully met with several government elites from all the pertinent entities, the size of this group is less represented when compared to the private stakeholders or R&D stakeholders' groups. This is mainly due to the fact that only a few government entities concerned with solar energy exist. Further, the members of this group were less open to having an interview in comparison to the other stakeholders' groups. While the private stakeholders' group offered valuable insights into the energy market status, both the public and R&D stakeholders' groups provided the author with a higher level of analytical expertise about the overall context

[4]During the time the interviews were conducted, the name of the Ministry of Energy used to be named as "Ministry of Energy, Industry and Mineral Resources".

of solar energy policies in Saudi Arabia. Thus, the population and distribution of the interviewees can arguably inform the studied case in a way that would lead to adequate and saturated data.

5.6 Results and Discussion

5.6.1 Landscape (Macro-Level)

5.6.1.1 Political System

SQ1. To what extent do the political system and the non-existence of independent Civil Society Organizations allow for system innovations in the solar energy sector, particularly bottom-up innovations?

The structure of political systems is critical for any technological transition in any country. The political system in Saudi Arabia is a monarchy. While similar patterns of the ruling are followed by countries such as the UK, Japan, Spain, and Belgium, Saudi Arabia, however, tends to be more of a combination of 'Oligarchy' and 'Democracy' (Al-Turaiqi 2008), or "globalizing monarchy" (Henry and Springborg 2001). Among the government, democratic practices are formulated in two forms. First, the Shura Council (Consultative Assembly), which is the formal advisory of the country that has the authority to interpret and draft, but not to pass, laws to the King and his cabinet and advise the King on regulations he submits to it. The Shura Council also has the power to call in the ministers of questioning and review the country's annual budget. Second, the state's three authorities (The Judicial Authority, The Executive Authority, and The Regulatory Authority) exercise its duties independently from each other. All the laws forced by all the governments must comply with the rules of Islam religion, from which the Saudi political system derives its authority.

Regarding the involvement of the society, although there are many Civil Society Organizations (CSOs), most of them are government-affiliated. In general, the CSOs sector is undeveloped, and its contribution to the general order is still limited. This is due to restrictive legal frameworks that impede such organizations to form and operate. A private stakeholder stressed on the restrictive regulations that cause the stumble of CSOs creation.

> Currently, there is no existence of solar energy independent CSOs; however, there have been many endeavors to establish them. We tried to establish a non-profit organization named "Solar Energy Association" to help increase the awareness about solar energy into the mainstream society and help in job matching of the solar energy industry, but so far we could not go forward with it because of the government bureaucracy. (Interview no. 15 – Private stakeholders).

Despite the Saudi cultural heritage, which includes the religious principles that call for civic work in different aspects of daily life, the restrictive environment of CSOs remains evident. In regard to CSOs relating to solar energy, the Saudi Arabia Solar Industry Association (SASIA) is a notable exception. It is arguably the only CSO that can be pointed at if we excluded other CSOs that are government-affiliated such as the Energy, Industry, and Mineral Resources Committee of Council of Saudi Chambers or those which operate both locally and regionally such as, Middle East Solar Industry Association (MESIA). SASIA introduces itself as "[…] *an industry promotional organization that brings together the national and regional solar industry.*" (SASIA 2011). Members of SASIA are drawn from all sectors of the solar industry, including PV, CPV, CSP, and solar thermal. The membership of SASIA is also offered to raw materials and component suppliers to the solar industry, such as glass, steel, and engineering corporates.

However, SASIA has been receiving criticism over its weak presence and contribution to the solar industry. The announcements of the SASIA's representatives participating in local and regional solar energy-related events have been limited.[5] A solar energy local developer confirms this notion:

> Three or four years ago, the government allowed for the establishment of new solar energy CSOs, but members of most of these CSOs are working to serve their personal needs instead of serving the profession or the industry of solar energy. (Interview no. 6 – Private stakeholders).

The significance of the role that the CSOs can play in the modeling and formulation of policies in a state might not be fully realized in Saudi Arabia. The weak realization of this culture may be attributed to the nature of the relationship between the government and its citizens. The top-down implicit social contact in the Saudi state, as well as other GCC states, is an evident characteristic of this relation (Kamrava et al. 2016). Thus, citizens tend to refrain from involving in civic activities that serve the common good as they assume that this is entirely the government's duty to fulfill. One local developer confirmed this notion by stating:

> Of course! The political system affects the system innovations…let's talk frankly. When the political system changed from the old one to a new one led by Prince Mohammad, as a young man, from where did he start thinking? From solar energy…renewable energy…electric vehicles (EVs) which started to be implemented in Saudi Arabia this year; therefore, the political system plays a big role in this matter, especially if you are in monarchy… In a country like Saudi Arabia and other GCC states, things are going on a top-down structure rather than bottom-up. As long as you don't explain to the people what you want to do and then make them involved, no one will take the initiative to move forward and this is the shortcoming of this region. In contrast to other countries, such as Jordan, which is also a monarchy, but people at the lower levels are influencing the higher levels. Solar energy industry there is an example; they are at an advanced level. (Interview no. 8 – Private stakeholders).

In short, the absence of an active role played by independent CSOs concerned with solar energy in Saudi Arabia has to be accepted as a fact. It is hard to expect a

---

[5]SASIA's official website has been unavailable to access and its social networking service (SNS) platforms have been inactive for over three years. In comparison to SASIA, MESIA which is founded by eight large companies, has been by far more active in organizing networking opportunities for solar energy professionals.

bottom-up system innovation from non-government entities in a power-centralized country like Saudi Arabia, where the contribution of the non-profit sector to GPD is merely estimated at 0.2% and planned to increase to 0.6% by 2020 (NTP 2016). Although most of the interviewees agreed that the political buy-in into solar energy would be the driving force for system innovations, the majority of the interviewees see no significant influence on the contribution of the independent CSOs' bottom-up innovations.

SQ2. How likely is the Saudi Vision 2030 goals for solar energy to be altered by changes in the political system?

Since the 1960s, Saudi Arabia has realized the critical situation of depending on depleted oil resources as the only national income. Over the years, Saudi Arabia developed what known as (Five-year-Plan), which aimed to diversify its economic income resources and increase the contribution of other sectors to the gross domestic product (GDP) were the top priorities. Nevertheless, no notable progress has been made (Amen 2017). Further endeavors were made by K.A.CARE to develop the renewable energy sector between 2010 and 2016; however, the set objective could not be met. The rapid population growth, coupled with the surging local consumption of fossil fuel resources encouraged the Saudi government to plan and launch a national economic reforming plan (i.e., Saudi Vision 2030). The primary objective of this grand plan is to move to the end of what the Saudi Crown Prince Mohammed bin Salman (then: Deputy Crown Prince) described it as "addiction to oil" (Reuters 2016; Aawsat 2016).

In 2017, the MOE launched the National Renewable Energy Program (NREP), a strategic initiative that operates under Saudi Vision 2030, aiming to maximize the potential of renewable energy in Saudi Arabia. This initiative is led by REPDO which was founded to facilitate the implementation of renewable energy projects across the country. REPDO set renewable energy planned capacity of a 5-year target (by the year of 2023) as (solar PV: 5.9 GW, Wind: 2.4 GW, CSP: 1.2 GW) and a 12-year target (by the year of 2030) as (solar PV: 40 GW, Wind: 16 GW, CSP: 2.7 GW). Early in 2019, these set 5-year-goal was later increased to become (solar PV: 20 GW, Wind: 7 GW) by 2024 (Fig. 2.1). These revisions of the renewable targets have been made without any particular change in the political system. A total of 70% of these projects is planned to be awarded to the PIF to negotiate deals with international partners while the remaining 30% will be tendered by REPDO.

The long-term Saudi Vision 2030 blueprint is designed and adopted by the Crown Prince Mohammad bin Salman setting long-term goals and expectations. Although this fact may draw the conclusion that the Saudi Vision 2030 and its goals are strongly related to the status of Prince Mohammad, this is may not necessarily the case. Over the last 30 years, no major discoveries of oil reserves have been made in Saudi Arabia or anywhere else in the world. Saudi Arabia, which used to depend on traditional small-scale trade and agriculture scattered across the Arabian Peninsula with a small population and low demand for energy, has transformed into an energy superpower with an increasing challenge to meet both domestic and international demand for power. One Public stakeholder stated:

> The government work is institutional, not only related to certain individuals. We are an oil-based country and all that oil will eventually deplete. So, alternatives need to be found to meet the demand for energy, job creation, and technology localization. In the past, when the oil prices reach USD100, the research and development of solar energy in Saudi Arabia would decrease, and if the prices go down the research and development would increase. But even now, despite how high or low the oil price is, even China and India continue to invest in solar energy. (Interview no. 5 – Public stakeholders).

Dependence on fossil fuel exportation as the only national income might have been worked in the past, yet the rapid increase in the infrastructures in line with population growth rang the alarm bells of the Saudi economy to take action to avoid an unpleasant future. The situation was critical that made some experts, in Citigroup for instance, predicted that if the local oil consumption keeps growing along with peak power demand, Saudi Arabia could even become a net oil importer by 2030 (Citigroup 2012). The majority of the interviewees confirm this notion. One interviewee representing an independent think tank stated:

> Prince Mohammad bin Salam is indeed the one who initiated the changing vision, but I believe it is an integrated institutional work in which we started to see in 2016 the creation and integration of the entire ecosystem. In the beginning, the vision was not very clear, but now we started gradually to see it. I think any change in the political system will not greatly affect the Kingdom's approach of the Vision 2030 or the projects of solar energy, as the public now accepts these projects. Solar energy is a reality that we must live with. (Interview no. 25 – R&D stakeholders).

In brief, similar to any newly elected government, which usually comes to power with a new vision, policies and regulations may change according to the new government's views. This also applies to the case of Saudi Arabia. That being said, it is important to emphasize that the solar energy goals are now an essential component of the Vision 2030, by which the government hopes to relieve the pressure on the local oil consumption and to diversify its economic incomes. Therefore, although the likelihood of solar energy goals to be altered remains high, the changes of these goals, if ever took place, may not be significant.

SQ3. To what extent are the formal decision-making processes the actual decision-making structures?

As outlined above, Saudi Arabia is slightly different from other monarchies, which can be described as a combination of 'Oligarchy' and 'Democracy.' The fuzzy overlapping nature of decision-making structure in Saudi Arabia has significant implications for policy forming and implementations, and thus for the capability of implementing solar energy policies. Until recently, government entities, to a certain level, used to work on a horizontal level reporting upwards to the ruling level, who eventually authorizes policies and makes decisions. The weak horizontal coordination between various government institutions and entities made the implementation of new policies a dragging task impeded by bureaucrats who have what so-call "veto power" that can even subvert the implementation of policies authorized by the ruling elites (Hertog 2010). To reduce the entrenched bureaucracy and to increase the performance efficiency of the government's institutions, twelve councils and committees were abolished and replaced with two new councils established by King Salman and led by his son, the Crown Prince. These councils are the Council of Economic and Development Affairs and the Council of Political and Security. After the formation of Vision 2030, the Council of Economic and Development Affairs, which included 21 minister and other high-level officials, developed an integrated governance system; to ensure the institutionalization of work and raise efficiency and facilitate the coordination of efforts between the relevant authorities to meet the objectives of the Vision 2030.

The decision-making process within the government entities has seen a remarkable improvement. This includes the solar energy policies decision-making process; a public stakeholder elite described this improvement as:

> The decision-making process has changed significantly. In the past, it was taking so much time due to the bureaucracy and committees' delaying. Proposed plans used to be submitted to the Minister (of Energy), where he does the filtering and send them back to the high committee, the Crown Prince and the King. But now, there is a higher committee for initiatives operating above all these high committees, the Council of Economic and Development Affairs, which led by the Crown Prince and the respective minister as a member of the council would be there to present the proposed plan. Even the set deadline strategy for each initiative has worked successfully. The reporting to the Council of Economic and Development Affairs is now received from only one reference body, in our case the MEIM[6]. (Interview no. 29 – Public stakeholders).

Another R&D stakeholder confirmed that by saying:

> The current situation is better than before because, in the past, the main energy-related entities were very segregated. There was not a strong alignment between us and other entities. We used to work separately. This also was true for SEC, ECRA and Aramco. The private sector was utterly isolated. Even the most essential thing, 'standards and regulations,' did not exist. When the Vision 2030 was formed, things started to be clear then the entire energy ecosystem undergone through a comprehensive restriction process and 11 entities were included under the MOE that was created lately. One of the useful measures applied is what so-called 'join-committee' where a representative from each entity meets every month to report their updates on the implemented projects. And there is also what so-called 'Steering committee' led by Prince Abdulaziz bin Salman to follow up on projects. (Interview no. 20 – R&D stakeholders).

While it is elusive to give one specific flowchart of the decision-making structure for the renewable energy systems in Saudi Arabia, based on the outcomes of the interviewed elites, the sequence below can arguably describe the general cases of how solar energy plans are developed:

- MOE works jointly with the relevant stakeholders such as analysis the market and develop scenarios of demand for energy and identify the needs

[6]In August 2019 a royal decree was issued to divide the Ministry of Industry and Mineral Resources into two ministries: Ministry of Energy (MOE) and Ministry of Industry and Mineral Resources (MIM). Currently the Ministry of Energy is headed by Prince Abdulaziz bin Salman, the King son's and the older half-brother to the Crown Prince Mohammad bin Salman. While the Ministry of Industry and Mineral Resources is leaded by the minister Bandar Alkhorayef, a former private sector investor and executive.

- A draft plan is developed and shared with relative entities to verify its soundness in both technical and economic aspects
- The draft plan is approved by the MOE' in-house and out-of-house consultants
- The minister of MOE approves the draft plan
- The draft plan is presented at the Council of Economic and Development Affairs by the minister of MOE
- The draft plan is turned to the Bureau of Experts at the Council of Ministers to be studied
- The draft plan is finalized and turned to the Council of Ministers for final approval
- The finalized plan is approved or returned to the Bureau of Experts at the Council of Ministers or to the relevant entity to be further studied.

The sequence mentioned above of the decision-making may vary based on the case being developed. It is important to note that since the formation of the Vision 2030, dynamic changes have been made to the structure of the government entities. This includes merging or splitting two ministries, pointing new ministers, and establishing new entities or departments and assign specific tasks to accomplish. The figure below represents the interaction map of renewable energy stakeholders in Saudi Arabia (Fig. 5.4).

Since the solar energy industry is still at an early development stage, more time seems to be required to codify intergovernmental cooperation on solar energy projects. A Public stakeholder elite described the followed measures to study one site before it is selected to build a solar park.

> It takes about a year to do a complete site study. Let us say we are selecting one area to study. First of all, we conduct typography, geotechnical, hydrology tests to confirm whether the site can be used or not. Next, we do resource measurement, such as solar radiation measurement, etc. After that, we do an environmental impact study. And then, of course, we conduct a grid impact study. How is the grid? And what is the number of MW to be installed? What is the maximum the grid can handle? [...] sometimes we say 5 or 6 GW it depends, some of these parks get canceled or cannot be done, for many reasons, for example, the Minister of Housing is planning to take over the land. There are many reasons like that. (Interview no. 3 – Public stakeholders).

However, this is not always the case. Some of the large-scale projects are still directly ordered by the top-level ruling leaders, as confirmed by many interviewees. A Public stakeholder elite underlined:

> I believe the formal decision-making processes match the actual decision-making structures by 90%. The remaining 10% are initiatives coming directly from the top. There is support at all levels. Sometimes a decision is made and when it is presented at the Council of Economic and Development Affairs, they return it to us saying: 'expand the scope of the project. Don't

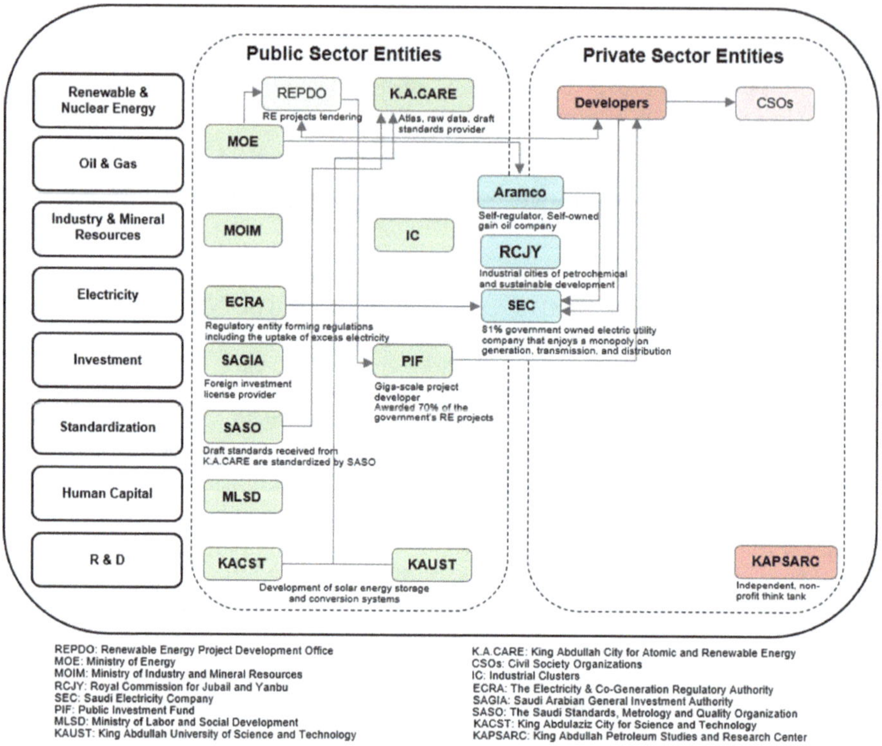

Fig. 5.4 Renewable energy stakeholders in Saudi Arabia (2019). *Source* Own compilation

> be conservative'. Good examples of this are Neom city,[7] the Red Sea Project,[8] Qiddiya.[9] That was only the leadership[10] behind it. There was a political desire that saw in these projects a chance by involving more than one sector. (Interview no. 3 – Public stakeholders).

In sum, it was demonstrated that most influence in the Saudi Arabia energy system lies with the entrenched regime actors, particularly in the political domain. Recent reforms within the government's both the horizontal and vertical decision-making

---

[7]Neom is a cross-border smart city in northwestern Saudi Arabia announced to be developed by the Crown Prince Mohammad bin Salman in 2017. This giga-project will incorporate technologies and also function as a tourist destination.

[8]The Red Sea is a megaproject announced in 2017, to develop as a luxury international tourism destination on the Red sea the western of Saudi Arabia.

[9]Qiddiya is an entertainment megaproject to be established in the capital of Riyadh and owned by the PIF. The construction of the project started early in 20 19 and phase one is expected to open by 2023.

[10]The term "leadership" is a common term usually used in Saudi Arabia and other GCC countries to refer to the top-level political ruling leaders.

processes have helped to match, to a high degree, the actual decision-making structure. However, further reforms are needed to improve the cross-cutting cooperation between the related government entities in order to increase the reliability and feasibility of solar energy projects to be implemented. Saudi Arabia requires all sites to be pre-approved by the government before it is open for tender. However, as can be understood, the identification of potential sites and developing a rough technical concept requires a long time to complete. This dramatically affects the speed of executing solar energy projects. Despite the large-scale reforms made covering almost every sector in Saudi Arabia, the existence of a centralized, top-down decision-making process is yet still evident.

5.6.1.2 Key Transregional Energy Governance Bodies

SQ4. Which role do transregional energy governance bodies such as OPEC, OAPEC, or IRENA play in the promotion or stalling of national solar energy policy in Saudi Arabia?

Transregional energy governance bodies have a substantial influence on policymaking. They help member states by providing robust analysis and coherent proposals and also help co-designing policies with the member states. This highlights the significant role the transregional energy governance bodies can play as part of the landscape-level. As a member of the three transregional energy governance bodies mentioned above, Saudi Arabia works jointly to tackle issues regarding the carbon footprints and oil price fluctuations. However, how significant is the influence of transregional energy governance bodies on solar energy policies in Saudi Arabia? A solar energy leading developer in Saudi Arabia who used to work for IRENA said:

> I used to work with them, I can tell you a bit about them. Basically, I'd not say they would influence the national solar energy of Saudi Arabia. IRENA is an international agency and Saudi Arabia is a member of it) and they do studies all over the world to show the real potential of renewable energy. And in some specific region, they work as advisors to a government to guide them on how to make the politics better, how to create new sectors, how to subsidize, how to enhance... and so on. But in any case, they don't have any power over the government. They don't come and say, OK... this is the way you should do it, and then they do it. No, all that they do is reports and recommendations to decision-makers. It's up to decision-makers in the country either to follow their recommendation or not. (Interview no. 17 – Private stakeholders).

Another R&D interviewee confirmed that IRENA has no influence over Saudi Arabia's solar energy policy nor does it on the solar energy policies of Abu Dhabi, the capital of the UAE, that hosts the headquarters of IRENA.

> These kinds of organizations are international. For example, does IRENA have an impact on the UAE specific other than the rest of the world? The answer is 'no.' The impact is political than technical. Hosting IRENA shows the willingness to progress technically. Masdar is focusing on renewable energy by default, not because IRENA is there. These transregional organizations have no impact at all on policies. They are trying to bring different views closer, enhance aliment, contributed to some countries that had some difficulties in identifying and

> setting their renewable energy goals and strategies. Saudi Arabia can do that, and Saudi Arabia designed the 9.5 GW[11] target by itself. (Interview no. 19 – Private stakeholders).

To summarize, OPEC and OAPEC are mainly reflecting the policies of the member states, and its activities are limited to oil trade in the upstream and downstream sectors of most Arab counties. Thus, it may have no impact on solar energy policies in Saudi Arabia. Unlike the European context, an intragovernmental organization such as IRENA seems as well to have no tangible influence over the solar energy policies in Saudi Arabia. IRENA may work closely with more entities in the hosting country, the UAE, such as Masdar City. However, it has been no evident influence on domestic renewable energy policies, which in many ways, would counteract the purpose of such transregional bodies.

5.6.1.3 Climate Change

SQ5. Is the climate change agenda actively considered in the political decision-making on renewable energy developments in Saudi Arabia?

As a massive oil-producing country, Saudi Arabia has exposed to a high level of GHG emissions that resulted from energy-related CO_2. GHG emissions in Saudi Arabia grew 75% between 1990 and 2013 (from 9 tCO_2/capita to over 16 tCO_2/capita) making it the second-highest (CO_2) per capita among the G20 member states. As a result, the Climate Change Performance Index (CCPI) ranked Saudi Arabia as a poor performing county comparison other G20 counties (Burck et al. 2015). Being the only oil-producing developing Arab country, it may not be a surprise that Saudi Arabia performs poorly at such an index. Another factor that should not be overlooked is the scarcity of freshwater resources. With a vast territory of 200 million square kilometers (km^2) characterized by hyper-arid climate and low rainfall levels, 25% of the country's oil and gas production is used by the cogeneration power desalination plant to produce fresh water and electricity (Badran et al. 2017). This is particularly is noteworthy because, since the formation of the Vision 2030, these areas have been a focus for policymakers.

In 1994, Saudi Arabia ratified the United Nations Framework Convention on Climate Change (UNFCCC) and followed that by ratifying and joining the Kyoto Protocol and the Carbon Sequestration and Leadership Forum (CSLF). The Saudi state's commitment to the environment protection can be seen in its active participation of all the UNFCCC negotiations and in its recently announced the strategic objective (No. 2.4.1: Reduce All Types of Pollution) under the National Transformation Program (NTP) (NTP 2016). However, is the climate change agenda actively considered when it comes to renewable energy development? One Public stakeholder stated:

[11] In early 2019, the 5-year target planned capacity of 9.5 GW was amended to become 27.3 GW (Turki Alshehri, 2019).

> This renewable energy development program of the Kingdom of Saudi Arabia KSA is not merely political decision-making on the climate change agenda. I believe that KSA's renewable energy program is backed up by several studies conducted by international agencies on the pollution caused by fossil fuel power generation and its impact on climate change and public health. I want to refer to the World Bank Report conducted on "Health Benefits of Controlling Air Pollution from Power Generation via Sustainable Energy Sources in Saudi Arabia" in May 2014. It was concluded in the report that the total estimated CO_2 emissions from Saudi Arabia power plants under the business as usual scenario, with no programs for introducing alternative energy to support the projected demand. The growth rate will increase from less than 150 million tons in 2010 to more than 441 million tons in 2030. These results suggest that unless renewables are introduced, pollutant emissions will increase, and the health of Saudi Arabians will suffer. On the other hand, by incorporating renewable sources of solar power to generate electricity to support the future demands, and by foregoing incorporation of crude- and diesel-based power plants, Saudi Arabia has the potential of reducing its particulate emissions 16% by the year 2020, and 43% by the year 2030. In terms of human health, these reductions in pollution could save USD 7 billion in health care costs and 2,500 lives annually. (Interview no. 4 – Public stakeholders).

Solar energy policies could interrelate with environment protection procedures in two different ways. First, utilizing the UNFCCC's Clean Development Mechanism (CDM) saleable credits which allow countries that ratified emission-reduction or emission-limitation to implement projects in developing countries. These credits could be sold to the member states of registered under the Kyoto protocol to secure the necessary funds to implement solar energy projects. Second, implementing utility-scale solar energy projects for seawater desalination could save the country's more than 80 million barrels of oil equivalent that is being consumed every year, as stated by the former governor of Saline Water Conversion Corporation (Alwasatnews 2015).

However, none of the interviewees raised any points neither regarding the CDM utilization as a potential monetary source nor the implementation of solar energy use for desalination. Several interviewees shared interesting perspectives about the whole concept of climate change, as one independent think tank interviewee put it "*As I work in this field, I can say, climate change is a marketing tool more than a real goal.*" (Interview no. 28 – R&D stakeholders).

A similar viewpoint was mentioned by a Private stakeholder who said:

> This is the first world's problem to fix, not ours. We should indeed be making the first step for solving this issue, but we're not taking the large part of the responsibility because the first world has been built on the shoulders of the third world. The U.S. and Europe have progressed at the expense of the third world and by plundering all its resources. It is good to be a part of the solutions, but we don't take responsibility for the large part of this issue, especially if we are not behind producing a large amount of CO_2 like China and the U.S. are. (Interview no. 7 – Private stakeholders).

In summation, while divergent opinions have been observed among the private and R&D stakeholders, all interviewed public stakeholders agreed that the government is considering the climate change agenda as a part of its commitments to different climate protection initiatives and organizations. However, it can be argued that the main driver of expanding or limiting the capacity of solar energy (and other renewable sources) is the necessity to meet the domestic growing demand for energy and the diversification of economic incomes. "*Climate protection is one goal among others*," as one of the top public stakeholder elite has put (Interview no. 1 – Public stakeholders). Government officials sounded cautious when it came to the topic of climate change. They seemed not to want to fight against the country's primary and vital source of income—fossil fuel.

5.6.1.4 Long-Term National Economic Diversification

SQ6. To what extent are oil prices related to the deployment of solar energy in Saudi Arabia?

As the world's largest oil exporter and the world's third-largest oil producer, the petroleum industry accounts for roughly 90% of export earnings, and 87% of budget revenue and 42% of GDP (OEC 2019). Since the 1970s, the government has been seeking to allocate oil revenues to transform other undeveloped sectors to build a more sustainable economy. Although most of the set goals for this purpose have not been met, the economy has grown significantly. Yet, the GDP remained subject to swing oil prices. Figure 1.1 illustrates the correlation between the crude oil price and the Saudi GDP growth over the last six decades.

It is noteworthy that policymakers in Saudi Arabia deal with solar energy with great caution. Several public and R&D stakeholders implied the pressure imposed aggressively by the oil-supporting lobby within the government, and Aramco was the main reason behind the stumbling of solar energy development in Saudi Arabia. However, after the fall of oil prices in 2014, have this perspective changed? An R&D stakeholder emphasized the differences between the context of Saudi Arabia with others by distinguishing between those motives in Saudi Arabia and other countries in pursuing to develop solar energy industries.

> In countries like Germany or Japan, oil prices are strongly related to the dissemination of solar energy. However, here (in Saudi Arabia) the government does not consider oil prices, but rather, the limited oil reserves that we still have. If you do not utilize this depleted resource smartly, it could affect the future of the government. Therefore, oil prices are not as important as the strategic selected choices that help preserve the resources of the national incomes of this country. (Interview no. 30 – R&D stakeholders).

Another private stakeholder from a leading solar energy developing firm described the effect of oil price on the development of a local solar energy industry as:

> I would say the low oil prices may drive the people to think to develop another form of energy. However, the higher prices would not prevent countries from continuing to develop solar energy sectors. So, in my opinion, in both cases, whether oil price continues to increase, with is beneficial to the country to have more money to invest in solar energy, or in case of oil price goes down should not impact the developments of the solar energy negatively. It should encourage the country to move ahead because the economic necessity leaves us no other choice. (Interview no. 16 – Private stakeholders).

In a few words, High oil prices make solar energy a more viable and attractive option for the replacement of conventional fossil fuel generation internationally. On the other hand, oil saving domestically shall help in exporting more quantity in the international market. Arguably, oil prices are not the only driver for the deployment of solar energy in electricity generation in Saudi Arabia but rather, as put by one of the public stakeholder elite, "*secondary.*" Other factors direct the deployment of solar energy are high solar potential, reduction of Solar energy generation price, environment-friendly projects, creation of direct, indirect, and induced jobs, improvement in public health, diversification in power generation technologies, and increase in the local-content manufacturing industry.

5.6.1.5 Country Branding

SQ7. Is country branding seen as a motive in the spread of solar energy?

Recent reforms made by Saudi Arabia in terms of the government restructuration, tourism enhancement, women's rights, tax introduction, anti-terrorism, and anti-corruption campaigns, and foreign investment attraction have positively contributed to the Saudi brand image. Although Saudi Arabia has no unified national brand, consultancy Brand Finance, which assesses the world's nation brands based on three key criteria (goods and services, investment, and society), ranked the value of the Saudi national brand as 21st, compared with 22nd, on the previous year. This growth is estimated at 18%, while it ranked 30th on the FutureBrand Country Index (FCI 2019).

In 2017, the PIF announced the Neom futuristic city, which will be powered by renewable energy sources. In the following year, PIF with SoftBank Vision Fund announced its plan to develop the world's largest solar energy manufacturing project with a capacity to produce up to 150–200 GW by 2030. Such ambitious projects can undoubtedly increase the positive image of Saudi Arabia both domestically and internationally as a peace advocate, eco-friendly country. Many living proofs show how sustainability can positively contribute to a country's branding image. For instance, according to City Nation Place (2020), Costa Rica became among the top 10 most searched online countries worldwide for "renewable energy" where the search rate jumped by 858% after Costa Rica implemented its renewable energy policies. However, among the policymakers in Saudi Arabia, these projects are planned to be linked in enhancing the Saudi national brand image? The majority of the interviewees from all the groups agreed with the perspective that country branding can be seen as a motive for the implementation of solar energy projects. Interestingly, divergent opinions have been observed among government stakeholders. One stakeholder from the MOE stated:

> Currently, I think the reason (behind introducing solar energy in Saudi Arabia) is, in general, an economic one. The comparison between Saudi Arabia and the UAE is difficult– we have a 10 times larger population. Solar energy could be used as a marketing tool within the private sector, but as a state– no. (Interview no. 3 – Public stakeholders).

Interestingly, another stakeholder who belongs to the same government entity has a different point of view, saying:

> No—solar energy is a part of the brand. Saudi Arabia is toward renewable energy and toward diversifying the energy mix—solar energy is one of many brands that we build. Take Neom city as an example. It primarily focusses on solar energy. (Interview no. 1 – Public stakeholders).

Although the majority of the private and the R&D stakeholders see the enormous 150–200 GW solar project announced by PIF and SoftBank Vision Fund can be used as a promoting tool for the country branding image, other stakeholders hold a different point of view. A private stakeholder based in Riyadh stated:

> The Kingdom differs from the UAE. The use of solar energy in Saudi Arabia is coming from an economic perspective, as it saves the Kingdom building a new power plant because the peak load demand occurs at the same time with the high solar irradiation. The economy is the factor that influences decision-making, not marketing. The UAE is not rich in oil, and it only exists in a few Emirates. As an importer of oil and gas, it can be understood why the UAE invests in solar energy—it has only a few options. (Interview no. 15 – Public stakeholders).

To summarize, Saudi Arabia is the 12th largest electricity-producing country in the world and the largest power producer in the Middle East region. Switching from conventional resources to renewable energy is one of the major shifts of policies in the region. Clear policies for renewable energy implementation have been announced for the upcoming years. Nevertheless, the immense planned capacity announced to be developed until 2030, seems not to be taken by the government stakeholders as a promotion tool of the national brand image. The majority of the private and R&D stakeholders believe that the development of solar energy projects should be associated with country branding. However, the prevailing opinion among public stakeholders considered the economic necessity as the force behind developing solar energy projects. In their argument, the public stakeholders claim that being the world's largest oil exporter besides being the house of the two holy mosques of Islam, is what promotes and sustains the brand image of Saudi Arabia in a way no other elements can do.

5.6.2 Regime (Meso-Level)

5.6.2.1 Technological Regime

SQ8. To what extent power generation, transmission, and distribution infrastructure in Saudi Arabia are sufficient for a successful deployment of residential solar photovoltaic systems?

The current installed power capacity is estimated at 87.7 GW. Although some media agencies reported about the stumbling of the SoftBank immense solar project 150–200 GW (Casey 2019; pv magazine 2019), this project if ever successfully implemented, would change the entire energy ecosystem in the Saudi state. Further, it was indicated that RSPSs with a capacity range of 1 kW to 2 MW can be integrated into the public electricity distribution system (ECRA 2017).

Following are the maximum limits for RSPSs to be installed in the existing capacity of the public power grid:

- Solar PV installations shall not exceed 15% of the distribution transformer capacity in any area.
- The maximum aggregate capacity to be installed in the area of any electricity department shall be up to 5 MW.
- The aggregated installed capacity shall not exceed the 3% peak load of a preceding year.

In addition to capacity extension challenges, it is noteworthy that any excess electricity produced by individual households who installed RSPSs can only be sold back to the same electricity provider—SEC. This would lead to an unavoidable conflict of interest which will be reflected in the entire plans of integrating RSPSs. One interviewee describes how the current relationship between related stakeholders is fragile and how it cannot sustain the growth of the RSPSs integration.

> ECRA has issued the regulations of RSPSs, which also required modifications to be made to both Saudi transmission and distribution codes. Good start, but not enough and more work needs to be done. Let's assume that I installed an RSPS; excess electricity will be sold to SEC, so SEC is the buyer and the seller! And it is impossible for SEC to be the regulator as well! That's why more work needs to be done. ECRA was trying to take the lead, but SEC claims that 'I own the grid; thus, anything shall go into it, it should go through us.' Eventually, ECRA agreed and asked at least to set the guidelines and the policies for that. (Interview no. 12 – Public stakeholders).

One interviewee who specialized in the Saudi and GCC' utility grids stated:

> When we talk about the 200 GW project, we here need to change the whole electricity import and export system, the entire energy ecosystem. There is no doubt the utility grid will need to undergo large-scale enhancement, and I mean by that 'physical enhancement. Utility-scale solar power projects will be developed in remote areas, far from the cities. So, we will need a transmission network that is able to transmit electricity from the solar parks to the cities. Further, renewable energy, including solar energy, as known, is intermittent, so, we must have the software capability to ensure the management of the supply. And this requires smart gird applications such as demand-side management. (Interview no. 25 – R&D stakeholders).

In summary, many countries allow up to 30% integration of electricity generated from renewable resources to avoid any cause of fluctuation within the grid. Government stakeholders confirmed that the Saudi electricity grid can take up the same percent before a further capacity extension needs to take place. However, it is questionable that this would be the case if the 200 GW-solar project is included. If the enhancement of the utility grid does not keep up with the increasing capacity of solar electricity integrated with the grid, there is a high probability for RSPSs integration levels to remain limited. In December 2019, SEC awarded a project of installing 10 million smart meters for consumers by March 2023 (SEC 2019). This surely would boost the dissemination of RSPSs; however, there is a need for a plan that manages how the electricity generated from both the RSPSs and the utility-scale solar park should penetrate to the power grid.

SQ9. Is nuclear energy seen as a competitor of solar power production?

Nuclear power is utilized in many countries to increase diversity in power production and increase the reliability of the electricity system. Unlike the intermittent renewable energy resources, such as solar and wind energy, nuclear power is served as a baseload power source side by side with oil and gas supporting a basic permanent load over the year. In 2017, the royal decree No. 43309 approved the Saudi National Atomic Energy Project with four key components: Large Nuclear Reactors, System-integrated Modular Advanced Reactor (SMART), Nuclear Fuel Cycle and Regulation and Control (K.A.CARE 2017). This was followed by the establishment of an independent legal entity to follow up and achieve the commercial interests of the National Atomic Energy Project in the Kingdom under the name of the Nuclear Holding Company. The Kingdom has adopted pressurized light-water reactors as an ideal option for power-producing reactors for several reasons. These reasons include the ease of maintenance and the feasibility and effectiveness of producing clean energy (K.A.CARE 2017).

The overwhelming majority agree that the nuclear energy option is seen as a complement to the energy mix in Saudi Arabia. One interviewee who directly involved in the National Atomic Energy Project stated:

> In the short term, atomic energy is not considered as a competitor of solar energy. While in the long term, it does. As you know, almost 4 million barrels are being consumed locally every day for desalination and if we continue doing this, we will end up consuming our entire oil production by 2032. This is why measures such as increasing electricity tariff, investing in solar and nuclear power, finding alternative energy sources, were taken. Regarding nuclear power, I don't see much risk there because the safety standers and the R&D are now very high and better than before. The International Atomic Energy Agency (IAEA) sets minimum safety guidelines and ever a country that wishes to develop nuclear power reactors must establish a nuclear facilities safety board. (Interview no. 20 – R&D stakeholders).

Only very few interviewees voiced concerns about the high cost associated with the development of atomic energy plants, as stated by one elite interviewee:

> I support using nuclear technologies for medical or R&D purposes, but I am against using it to generate electricity. First of all, nuclear energy plants are prone to disasters, and if a technical error or a disaster occurs, it will wipe all benefits that we may have gained from establishing an atomic energy plant. Second of all, we need 20–25 years for training our youth to master this technology and be able to conduct plant maintenance. Third, in most cases, even among the developed countries, the overall cost of nuclear power plants is not precisely calculated. For instance, the cost of insurance, securing a location and the land's radioactive contamination clean-up is often not included. I am against nuclear power because we have another option, which is solar energy. Why do not make full use of solar energy and then see if there is a need for atomic energy? The Crown Prince led the signing of the 200 GW project. This would treble our need for energy. (Interview no. 9 – Private stakeholders).

Among the GCC states, the UAE took the lead and developed the first commercial nuclear power station named "Barakah" incorporation with Korea Electric Power Corporation. Following the stapes of the UAE, Saudi Arabia started to develop its nuclear program for peaceful purposes, seeing it as a complement of other energy forms. Unlike the UAE, Saudi Arabia aims to enrich uranium within Saudi Arabia, which has been a sticking point with the United States' so-called '123 agreement' that requires Saudi Arabia no to enrich uranium locally (Atomic Scientists 2019). Apart from the USA influence which, according to Bahgat (2006), caused nuclear development to delay, Saudi Arabia is also looking into other counties such as Russia to collaborate with to develop its nuclear power capability. Saudi Arabia has the ability to move forward with the 150–200 GW project and secure its future needs from electricity. While it has been apparent that nuclear option is not seen as a competitor of solar energy, it can be argued that the motives behind developing self-own nuclear capability may not be fully economic but also involves political implications.

5.6.2.2 User and Market Regime

SQ10. Is there any plan for residential solar photovoltaic systems promotion in place?

One of the primary components of the Vision 2030 is to eliminate subsidies from different types of public services that have been offered for several decades. For instance, in 2018, the electricity tariffs for the residential sector have been increased between 80 and 260%. Similar measures were applied to other public utility services such as water, petroleum, and gas and further reforms are to be imposed gradually until the tariffs meet the international standers. The rationale behind such measures is to restrain the rapidly growing demand for energy sources. As highlighted above, the residential sector in Saudi Arabia is accountable for over 50% of the total electricity consumption (Tlili 2015). This urges the decision-makers to enable individual households in order to possess RSPSs in their residences and integrate them into the public grid under the scheme on Net Metering. The interviewees' answers to the question of this section were divided deeply where one-third of them do not know if there are any RSPSs promotion plans in the place. Solar energy developers and firms' owners and executives criticized the opacity of this matter as the core of their business fundamentally depend on how the policymakers will proceed with this plan. One private stakeholder stated:

> No, there is no plan. Currently, the RSPSs are not subsidized, yet they want it to compete with other subsidized energy forms! This is not fair. If we wish to encourage people to adopt RSPSs, they should be subsidized like other fossil fuel resources that were subsidized. The payback period of the RSPSs is between 15 and 20 years! No one will be able to choose this option. (Interview no. 15 – Private stakeholders).

Interestingly, during conducting interviews, some of the interviewees who are directly involved in the RSPSs energy regulation asked the interviewer about updates on regulations, when they knew that the interviewer has talked with representatives from other entities. Further, a variation of the number reported by each entity of applications received from individual households to install RSPSs was observed. This may indicate weak coordination between the pertinent parties. One Public stakeholder clarifies the status of the RSPSs implementation as follows:

> I think the ads will increase gradually. They have started through newspapers and they will continue increasing. We are receiving inquiries, people are asking, and SEC received more than 2000 preliminary requests. The public awareness is essential. (Interview no. 2 – Public stakeholders).

In short, the stability of the public grid is required and excessive integration of RSPSs must be avoided. The pertinent entities seem to have set the necessary regulations to ensure all possible causes of grid fluctuation are eliminated. That being said, the number of preliminary applications required to install RSPSs received from individual households is significantly lower than what was expected. Concerning energy regulation in buildings in Saudi Arabia, it is more command-and-control energy regulation rather than market-driven energy regulation. The command-and-control energy regulations mainly rely on different levels of government entities rather than involving individuals (Sheng et al. 2018). For instance, activities of energy efficiency and energy conservation focus on public services-related buildings, such as government buildings, mosques, hospitals, and schools, while such measures are barely implemented in the residential buildings. As time passes, no notable progress has been made to attract more households to install RSPS technologies. It is unlikely for this situation to improve anytime soon under the Net Metering scheme currently introduced by ECRA, especially with the absence of identified goals and long-term strategy that link the RSPSs capacity with the national plans of renewable energy.

SQ11. Do you believe the transfer of European renewable energy policy design models is a viable method to spread residential solar electricity in Saudi Arabia, or would indigenous policy models work better?

Governments around the world have adopted various types of incentive schemes. Some of the schemes are still being used, others have been phased out. In 2018, Saudi Arabia established a framework for the Net Metering agreement to uptake excess electricity generated from RSPSs. This means eligible end-users receive credits for any surplus generation at the same price the purchase electricity from the distribution service provider (ECRA 2017). All the interviewees see that designing indigenous policy models is the most suitable option. One interviewee stated:

> We will develop our own policies. We cannot be like the Germans, because the Germans are paying the bill for the rest of the world. There are going to be lessons to learn. There are going to be a reverse of FiT, and quantity and price caps. The current Small-Scale Solar PV Systems Regulations issued by ECRA has a generation cap of 15% to avoid oversupply. I do not think the European examples are not applicable in our case. (Interview no. 27 – R&D stakeholders).

An interviewee who is directly involved with the formation of electricity regulations and polices confirmed that Saudi Arabia is not reinventing the wheel stating:

> Think global, act local. We have to consider our limitations, grasp the lesson learned, then design our own policies and set benchmarks our performance. (Interview no. 32 – R&D stakeholders).

The lack of the region-specific standards of solar energy forced Saudi Arabia to take the lead through KACST to develop what so-called "Desert Protocol," which is standards of solar PV panels specially designed for desert climate based on the International Electrotechnical Commission (IEC) standards. A senior consultant who is in charge of standardizations stressed on the tough negotiation they have had in order to adjust the solar panels codes fit the local desert climate.

> Besides the long negotiation we needed to modify the thermal cycling test standards, it took us three years of negotiation to get the approval to modify the solar panels code to bear heat from 70 °C up to 85 °C. Manufacturers have objected because this means additional incurred cost and the World Trade Organization (WTO) has resisted. However, after installing the PV panels for a while, we could prove to the WTO that panels would experience system failure if the code is limited to 70 °C and it should be increased to reach 85 °C. (Interview no. 21 – R&D stakeholders).

To summarize, there was a consensus by all the interviewees on the policy models of the RSPSs should be developed indigenously based on a well-understanding of the lessons learned from the global experiences. European renewable energy policy design models would be designed based on climatic conditions, circumstances, environment, tariff design, and electricity market structure of the European context. Apparently, it is well realized by the RSPSs policy designers in Saudi Arabia that foreign policies could not be copied and implemented, and local challenges imposed by the different climatic conditions and grid capacity limitations, and cultural differences seem to be taken seriously into consideration. Further, building on the IEC solar panel standards, KACST has recently developed the "Desert Protocol" which is a unique test specially designed to simulate desert conditions for solar PV panels. In comparison to other GCC states, Saudi Arabia, in this sense, is in an advanced position. Further, as confirmed by an informed stakeholder elite, other GCC states openly share their desire to adopt the Saudi regulations and standers once they are ready to use.

SQ12. What is the most significant factor that may act as a barrier to the deployment of residential solar systems?

A report released by the General Authority for Statistics indicates that only 1.45% of houses in Saudi Arabia are installing RSPSs, where Hail, northern Saud Arabia has the highest rate of 2.8%, and Bahah located in southern Saudi Arabia recorded the lowest percentage with 0.59%. Both cities are considered rural areas with partly developed infrastructure. The capital, Riyadh, recorded 1.78% of the houses to be using RSPSs to produce electricity (GAS 2019a, b). Since the regulatory framework was intended to come into play only in July 2018, it is hard to precisely determine the true causes of the significantly low number of houses installing RSPSs, although many insights can be gained from the interviewed stakeholders. Half of the interviewees see the economic and financial related factors are the most determinant while the other half divided between believing the culture and social awareness, and the political and market distortion factors the most determinant factors.

> I believe it is economic feasibility. If individuals can save money, they will be encouraged to install RSPSs. The solar panels and the aesthetics aspect are concerning point. (Interview no. 11 – Private stakeholders).

The head of REPDO describes the critical challenge of the spread of RSPSs as:

> I would say the alignment with all the government entities that are involved is the biggest challenge. Because you will have to conduct different studies regarding social aspects, environmental aspects, etc. We will have to align with all other organizations to make sure that this is what required for renewable energy. So, maybe this is the biggest challenge to make sure everybody is on the same page. (Interview no. 1 – Public stakeholders).

The integration of RSPSs in Saudi Arabia is still at the early development stage, and further efforts are required for a successful and smooth implementation. Besides the necessity for alignment between government stakeholders, economic viability appears as one of the most dominant factors of the RSPSs successful dissemination. RSPSs are spread in those countries wherein solar energy generation cost is less than the electricity generated by conventional means or where RSPSs adoption is encouraged by incentive instruments offered by those countries. At the utility-scale level, REPDO awarded the first IPP project to a Saudi Arabia-based joint venture at a world record tariff of $0.0234/kWh. At the residential level, the alignment between all the government entities, subsidized residential tariffs, and the RSPSs long payback period in Saudi Arabia are considered as significant barriers.

5.6.2.3 Socio-Cultural Regime

SQ13. How widespread is the knowledge of residential solar systems among individuals in Saudi Arabia?

This is a fresh area that has hardly been researched.[12] As confirmed by several public stakeholder interviewees, no official research regarding public awareness towards RSPSs has been conducted by the government entities (Al-Barrak 2018; Al-Taher 2018). Literature reviewing the status quo of renewable energy in the context of Saudi Arabia is mainly a technical-focused analysis. Some of few studied that investigated social perception towards RSPSs in Saudi Arabia, indicated that individuals have a good cognition level (Alrashoud and Tokimatsu 2019) and are in favor of the utilization of solar energy among other energy forms (Mosly and Makki 2018). Recent reforms of energy subsidies and the continued announcements of tendering new solar energy projects might have contributed to the public awareness towards RSPSs. This is coupled with the increasing number of registered solar energy developers and consultants, as confirmed by several interviewees. One interviewee made a remark indicating to "*an increase in the awareness about RSPSs.*" as follows:

> There is a movement, SNS platforms and the government's announcements also play a role in increasing the awareness about RSPSs. There are more than 200 contractors, developers,

[12] See a summary of the key previous studies on the interaction of individuals with renewable energy in Saudi Arabia in Table 1.1.

> and consultants who have registered as a certified service provider by SEC. As the electricity tariffs are increasing gradually, people started to talk more about RSPSs and its feasibility. And after people had started using energy-efficient appliances, LED, and thermal insulation in new houses, they started to notice the benefits of using such technologies which are reflected on RSPSs. (Interview no. 5 – Public stakeholders).

Unlike the large cities in Saudi Arabia, in rural areas where the electricity infrastructure is not fully developed, micro-scale solar PV systems can be seen in remote farms and Bedouin gathering. Modifications have been made to schools' curriculums, and topics about renewable energy, including solar energy, have been added. An interviewee who is a CEO of an investment and construction company stated:

> I believe we have a good level of knowledge about solar energy. Even my nephew came to me asking to help him with his school homework that was about the solar panels set on the rooftop of his school. (Interview no. 10 – Private stakeholders).

Most of the interviewees of all three categories believe that individuals have a good understanding of the basic concept of RSPSs. At the same time, the interviewees' expectations have dropped significantly; they were asked if the individuals have a good cognition level of the advantages/disadvantages, benefits, and system components. This may indicate the necessity to enhance the education programs and information provision for the individuals to develop a solid understanding of this technology, which is a crucial element for the successful diffusion of any innovations.

SQ14. Does religion play a role in promoting the use of solar energy technology?

Islam religion is the formal state faith of Saudi Arabia. The word Islam means peace, salvation, and submission to God. Saudi Arabia occupies a special place among Muslim countries as it is home to the Two Holy Mosques of Islam. Religion profoundly influences life patterns, attitudes, and everyday practices in Saudi society (Ayaz et al. 2003; Erhun 2001). In religious communities, religious congregations are often uniquely able to motivate people to make decisions to support the cause for which they mobilize.

In recent years, serval religious movements have been actively supporting the implementation of solar power generation units around the world. In the Philippines, Catholic leaders put pressure on the government to drop coal by working with activists and demand to install solar PV panels on church rooftops (Balea 2015). In India, the Gayatri Mantra, which descended from the ancient Rig Veda, the most revered of hymns in Hinduism, is dedicated to the ultimate source of energy: the sun. The Narendra Modi's government took advantage of this quality and sought support from the godmen of India to implement its ambitious solar energy program. The government asked the godmen to work as brand ambassadors promoting solar energy among millions of their followers and encourage them to install solar PV unites on their hermitages or ashrams (Manu 2016). Similarly, in Japan, several religious organizations, including Shinto shrines, Buddhist temples, and Christian churches, are installing solar power units at their facilities under the banner of the Religious and Scholarly Eco Initiative to confront environmental crises.

In Saudi Arabia, Islam is the state's formal religion. The word Islam literally means peace, salvation, and submission to God. Islam calls for moderation and stresses the significance of moderate use of natural resources protecting the environment (Ayaz et al. 2003; Erhun 2001), which may be used as an entry point for the promotion of RSPSs.

In religious communities, religious congregations are often uniquely able and motivated people to make decisions to support whatever cases called upon by those congregations. Responses to the question above divided equally and can be categorized into three groups. The first group agrees that RSPSs can be promoted by highlighting the principle of Islam that emphasizes on sobriety and environment protection. One interviewee stated:

> Yes, I think religion can play a role in promoting the use of solar energy because global warming and climate change are harmful, and religiously, everything that causes harm, like drinking alcohol, for instance, must be stopped. Therefore, religion can play a role in putting a limit on pollution and climate change. (Interview no. 23 – R&D stakeholders).

The second group was in the opinion that religion is somewhat likely to be used as a promoting tool. However, at the same time, they were not very optimistic about the influence the religion can have on the individuals' decision-making. While the third group sees no value of using the religion to inspire individuals to adopt RSPSs. Although the third group agrees that the teaching of Islam is compatible with the potential benefits gained from the spread of solar energy, they believe that approaches other than religion can be used to encourage RSPSs adoption. One interview highlighted this perception by saying:

> I do not think that religion can play any role in this regard. Our religion (Islam) indeed calls on rationalization in everything including electricity consumption and money use. However, I do not see any association with solar energy. For instance, solar energy can be promoted for being an unlimited source of energy, but I do not think this has to do anything with religion. (Interview no. 24 – R&D stakeholders).

In brief, contrasting opinions were observed among government stakeholders where some of them perceive religion as a potential promoting tool for RSPSs, while others do not believe that making use of religion would be an appropriate approach. Although the government has been using religious materials (e.g., verses of the Qur'an and Hadiths)[13] in campaigns targeting individuals to boost the rationalization of water consumption (Qatrah 2019), the results of this study revealed that there are no such established activities in the case of RSPSs. This was confirmed by several government stakeholders who are even in acceptance of using religion as a potential promoting approach.

[13]'The Qur'an' or Koran is the holy book of Islam religion, while 'Hadiths' is an Arabic word for "News" or "Story" and is record of the traditions or sayings of the Islamic Prophet Muhammad.

5.6.2.4 Policy Regime

SQ15. What is the residential solar systems' share of the total solar energy capacity to be installed by 2030?

The overwhelming majority of both the private and R&D stakeholders reported that they are not informed of such information. Several Solar energy local developers expressed their anxiety about the opacity and lacking transparency regarding the RSPSs share of the energy mix as the provision of such details would it strongly related to their existence in the market. Governmental stakeholders indicated that determining a specific number at this stage would be very difficult. Interviewees from both ECRA and SEC expressed the disappointment of the low number of applicants received from individual households to install an RSPS. One local developed stated:

> Frankly speaking, I do think it would not be a big share. If we can set something between 10 and 15% and accomplish it, it would be great! I believe that the lion's share will be for the utility-scale projects, which will attract businessmen once there are sure that the return of investment is high. [...] after one of our company's promo videos went viral, our call center received more than 250,000 calls within two months. There is a good level of awareness, but it needs to be increased as we are now at a transitional stage. (Interview no. 13 – Private stakeholders).

Another local developer emphasizes the critical role SEC can play in determining the success of failure of involving the individual stakeholders with the solar electricity generation by stating:

> It essentially depends on who serious SEC is to involve the individuals in the generation of solar electricity. If it worked at the beginning, more people would be encouraged to install RSPSs. But if the agreement between the two parties is not an equitable and unexpected issue arises, the individuals will be reluctant to move ahead and install RSPSs. (Interview no. 14 – Private stakeholders).

In summary, the lowest slab of residential consumer tariff was commenced from 5 Halalah[14] H/kWh (0.013 cents/kWh). Recently it has been raised to 18 H/kWh (0.047 cents/kWh) which is considered as a first step in removing energy subsidies on electricity. It is anticipated that subsidy would be removed gradually by the government until it meets the international standards. This measure would help solar energy to compete against other conventional energy sources. The recent increase in electricity tariff encouraged individuals to collect information about the RSPSs and examine their economic feasibility. Although ECRA set caps on small-scale solar PV penetration to avoid grid fluctuation, the results show that no specific target of planned capacity is set for the RSPSs.

SQ16. How seriously do the government bodies consider the social aspect when removing energy subsidies?

Social acceptance is an essential element of successful diffusion of any innovation; RSPSs is not an exception. As outlined earlier, the gradual abolishment of energy

[14]See Footnote No. 2.

subsidies in Saudi Arabia would help to make RSPSs a more attractive alternative for individual households against other electricity conventional generation methods. Nevertheless, the cancellation of energy subsidies may discompose the implicit social contract, which unlike the European countries, remains a focal point that cannot be overlooked in the hydrocarbon-rich GCC states (Kamrava et al. 2016). For that, it is critical to understand how the government entities in Saudi Arabia, which are keen to develop a diversified economy, and secure long-term sustainable growth, consider the social aspect when removing energy subsidies. Based on the results, this area appears to be considered carefully by the top decision-making rulers, as confirmed by the overwhelming majority of the interviewed stakeholders who emphasized the uniqueness of the political system in Saudi Arabia. One interviewee stated:

> Let us not forget that when it comes to the energy provision in Saudi Arabia, the political relationship here differs from the rest of the world. Do not forget that the government subsidies have been there for a long time. So, it is difficult to change the entire energy market at once. You need to design unique polices to make this transition smooth and peaceful. Thus, I believe foreign policies, whether they have succeeded or failed, do not apply to the Saudi market whatsoever. The Saudi energy market, in particular, is very different from the others, and there may be serious political and economic consequences if policies are not designed appropriately." (Interview no. 28 – R&D stakeholders).

Several stakeholders confirmed that the correction of the energy tariffs is studied very carefully from different aspects to minimize the negative impact of GDP growth rates and maintain the country's political stability. At this point, no subsidy provision program to enhance the deployment of RSPSs exists, and for the present time, the Net Metering scheme remains the only option for those who wish to install small-scale solar PV systems. Unlike some of the GCC states, Saudi Arabia, for decades, has been providing low-priced utility services to both citizens and expatriates at the same rates. Additionally, until 2017, energy subsidies were inequitably distributed with no fair caps or segmentation of users, which allowed high-income households and energy-intensive industries to take the most advantage than other users. Gradual correction of energy rates has been and will continue to be applied until it meets the international standers, which eventually is expected to result in an overall favorable effect on the welfare of households (Blazquez et al. 2017; Gonand et al. 2019).

> Definitely, social impact is a critical factor. What is the main driver for all these corrections carried out now? It is to correct the balance of subsidy provision. The primary mission of government officials is to support their citizens. But you ended up supporting everyone. And everything that you export became subsidized. The government has been subsidizing basic food products for particular reasons, also some industry sectors, manufactories, and electricity, here you see also some directed and focused subsidization which the way it should be. The social factor is taken into consideration, to prove that, we were able to wave all the subsidies at once, but no, you have to have a gradual approach to let the people adapt. The Fiscal Balance Program[15] is a good example. (Interview no. 19 – R&D stakeholders).

[15]The Fiscal Balance Program is one of the basic programs to achieve the 2030 vision which aims to strengthen financial management, restructure the financial situation of the Kingdom and create various mechanisms for reviewing revenues, expenditures, various projects, and the mechanism for their approval (FBP 2017).

To conclude, the uniqueness of the Saudi citizen-state relation is being considered by the high-level government stakeholders with great seriousness when removing energy subsidies. However, under the ongoing economic reforms and the abolishment of subsidies, there were no clear answers to whether subsidies should be offered to households and how these subsidies should be afforded without undermining the government's efforts to minimize the concept of the "*rentier state*". This has a great implication of the political economy in Saudi Arabia, especially with the absence of a clear vision that roadmaps the penetration of RSPSs.

5.6.2.5 Science Regime

SQ17. Is there an attempt to create a domestic knowledge hub for renewable energy technologies, and is there an industry structure that supports the spread of technologies into mainstream society?

This subsidiary question aims to understand if there is any intention to create a local knowledge hub that attracts brains, connects experts in a similar role across the country, and provides them with access to a wealth of knowledge to make better use of renewable energy resources. It also attempts to answer if there is an industry structure to disseminate technologies into mainstream society. While one third on the interviews had no information on this matter, the majority of the remained interviewees pointed at KACST as the primary knowledge hub existing in Saudi Arabia with no attempts to establish a new body. Some obstacles were identified based on the interviews, including the overlapping work between a number of the pertinent entities. Several interviewees embrace the efforts that were recently made to improve the decision-making structure of renewable energy in Saudi Arabia. Yet, at the same time, they criticize scattered efforts and the overlapping between some entities claiming that it impedes the development of the renewable energy industry. An ex-K.A.CARE director and a renewable energy projects director in a large private company stated:

> It is scattered efforts. Every entity claims they own it. It was supposed to be K.A.CARE, but now the directions have altered, KACST was originally that one to look at. All of this has changed with the new Vision 2030 when the MOE assigned each entity with specific tasks based on their strong points. Research and development of renewable energy technologies are assigned for KACST, technology localization and value chain creation for K.A.CARE, regulatory frameworks for ECRA and electricity receiver and distributor to SEC. (Interview no. 12 – Private stakeholders).

Although after the recent restructuring of the energy sector that allowed each entity to operate based on its core high-value specialty, several interviewees pointed to an overlap between KACST K.A.CARE, ECRA, and SCE in areas including regulatory requirements, standardizations, and public awareness enhancement. An elite interviewee stated:

> As for now, there is no industry structure that supports the deployment of renewable energy technologies into mainstream society; we are still at the early stages. But with regard to the

> knowledge hubs, they have come a long way. We may see a merge between KACST and the K.A.CARE's Renewable Energy and the MOE' Renewable Energy Center. It is easy to create a joint-unit dedicated to renewable energy because the infrastructure is there. (Interview no. 29 – R&D stakeholders).

The majority of interviewees who sees KACST as the main knowledge hub voiced concerned that after the announcement of the 150–200 GW major solar project, more knowledge hubs need to be established to cope with immense capacity expected to be installed and to minimize the gap between the R&D and industry sector. This may also indicate the lack of interlinking relationships between the knowledge centers and the industry structure. A general director of a renewable energy data center stated:

> We are looking for much more as a data reference: specialized people, decision making, researcher or public to have knowledge portal tools residential solar system where individuals can go and learn. (Interview no. 19 – R&D stakeholders).

Comparing to other Arab countries, Saudi Arabia ranked first in the number of scientific publications with more than 14,000 papers (Fig. 5.5). In 2013, Saudi Arabia achieved a 373% growth in output, which is considered the highest rate of growth during the last decade among the G20 members. Most of this growth has been since 2009 (Al-Suwaiyel 2015). With a total of $1.8 billion in expenditure on R&D, Saudi Arabia ranks 37th internationally, yet it is still low compared to other G20 member countries (qafilah 2016).

R&D stakeholders indicate the lack of funds needed to push renewable energy development forward. One research associate of an independent think tank stated:

> The industry structure is still missing, and to solve this, we need to secure funds to develop a certain technology, which is available but needs to be increased. We hope if there is a platform that accepts and examines research to validate its economic and technical visibility.

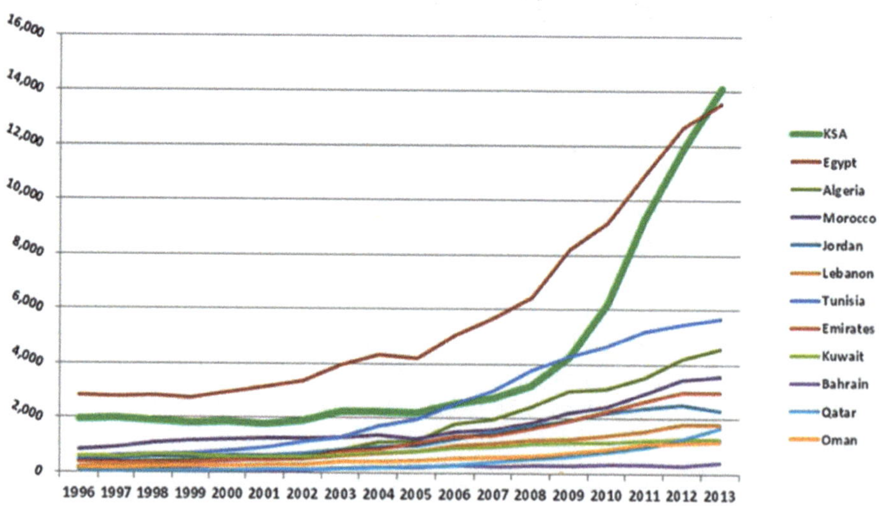

Fig. 5.5 Numbers of scientific publications in Arab countries. *Source* Al-Suwaiyel (2015)

> The private sector can fulfill this role as well, but the R&D sector in Saudi Arabia is more developed. Renewable energy is still immature in Saudi Arabia which will need us a long time to get there. This is also true of the industry sector. (Interview no. 25 – R&D stakeholders).

In summary, there is no clear vision to establish a new knowledge hub of renewable energy anytime soon. Further, although there is a movement within the MOE to improve the alignment between the respective entities, conflict of interest at managerial levels, including regulatory requirements, standardizations, and public awareness enhancement are appeared to be hindering the development of the renewable energy industry. In addition, the weak interlink between the R&D and the industry sector and lack of funds for the R&D sector constitute barriers to develop a solid domestic renewable energy sector.

5.6.3 Niche-Level (Micro)

5.6.3.1 Technological Developments

SQ18. What is your opinion on technology choices and the potential of the solar energy industry in Saudi Arabia?

The average sunshine in Saudi Arabia ranges between 8 and 9 h a day. The country receives one of the highest solar radiations of 2050–2450 kWh/m^2 yearly, which can generate 732–952 TWh in a year through Solar PV technology. Radiation is higher during summer, conveniently when the country's demand for electricity is higher and lower in winter when the demand is also substantially lower. Also, the hours of maximum output from solar units coincide with the peak hours of the day in the summer (12:00 P.M. to 5:00 P.M.) (Almasoud and Gandayh 2015). REPDO set renewable energy planned capacity of the 5-year target (by the year of 2023) as (solar PV: 20 GW, Wind: 7 GW) and 12-year target (by the year of 2030) as (solar PV: 40 GW, Wind: 16 GW, CSP: 2.7 GW) (Fig. 2.1). The majority of the interviewees were of the opinion that solar PV technology overweight CSP technology in terms of cost-effectiveness and the ease of applicability. A consultant from Saudi Aramco emphasizes that despite the advantages that CSP technology has as being a dispatchable power with less fluctuation, solar PV technology still a more suitable option for Saudi Arabia. The consultant continues:

> We have several excellent small-scale solar PV manufacturers that provide an excellent guarantee, but they need to be supported. Unlike PV technology, CSP technology requires large-scale investment and the economic feasibility of PV technology is higher than the CSP's. (Interview no. 18 – private stakeholders).

Another elite interviewee from KACST who also agrees on the high economic feasibility of the PV technology sees political stability is essential to develop a solid solar energy industry by stating:

> […] all options are there, but as for silicon-based solar cells in Saudi Arabia we hold the entire value chain. The potential is high, but the political and security stability first, then the economic stability second are essential factors. Ten years ago, we did not have the knowledge necessary to manufacture solar panels. In 2010, a factory was established within KACST by Saudis with the capability of manufacturing 3 MW. Today it manufactures 100 MW. Since KACST is a non-profit entity that promotes technology localization supports investors of the industry sector, several investors had been invited to visit this factory as an example of what a solar PV manufacturer looks like. They were surprised by the simplicity of the technology involved to establish such a manufacturer. Now we have 4–5 serious investors who are willing to establish new factories of their own, and we expected more after the announcement of the grand 150–200 GW solar energy project. (Interview no. 31 – R&D stakeholders).

Due to the increasing local demand for energy, the government shows a firm commitment to diversify its power sources, and solar energy, as highlighted earlier, is supported by the political buy-in. This political buy-in did not exist at least four years ago. On interviewee:

> In the past, developing solar energy projects was something like as if you work against the government's agenda—all of this has changed now. (Interview no. 22 – R&D stakeholders).

In brief, both solar PV and CSP technologies are encompassed in the government's renewable energy plan. More opinions were in favor of PV technology utility-scale projects over other options due to its cost-effectiveness and. Stakeholder elites from the MOE confirmed that overall targets for each renewable energy source are ultimately approved by the "leadership"[16] and then the technology choices are left to the private sector as the new role of MOE is limited to create a competitive ecosystem for the solar energy project developers to compete against each other and bring the best technology available.[17] MOE aims to achieve this by awarding the developers IPP projects and enhancing transparency, R&D, inventions and ensure that the developers follow the Saudi standards allocated for each form of energy and that they hold the proper expertise and qualifications.

SQ19. How prepared is the country to develop a robust Local Content in the solar power sector?

In 2019, REPDO announced tendering 12 projects for competition in as a second phase of the NREP with a total capacity of more than 3,000 MW. Six projects were launched in July and August, while the remaining six projects are planned to be tendered in the last quarter of 2019. The projects fall into two categories: (Class A), under which projects with a capacity of 100 MW and below, and (Class B), which includes projects with a capacity of more than 100 MW. The reason for this division is, according to REPDO, to create more opportunities for the local companies to participate in the national renewable energy program. In February 2019, applications

[16]See Footnote No. 18.

[17]In 2018, NREP launched a public database of more than 100 local companies operating in the renewable energy sector (wesupplyrenewables.com.sa). This portal aims to encourage private sector investment and build confidence in the renewable energy sector by offering domestic suppliers an access to a multi-billion-dollar renewable energy investment programs.

were announced for the first batch of renewable energy projects for the current year 2019, with a capacity ranging from 20 to 600 MW, where 60 companies, including 28 Saudi companies, were selected as qualified. This figure is the largest in terms of the number of companies that were previously selected (42 companies) in the first phase of 2017. REPDO sets a target to increase local content in the industrial and service value chains and expertise in the renewable energy sector between 25 and 35% (REPDO 2017). Although half of the interviewees consider Saudi Arabia ready to develop a robust local content of the solar energy technologies, the other half voiced concern. The pessimists pointed to lack of workforce and expertise, arguing that Saudi Arabia, for decades, could not localize the oil and gas technologies which fuels speculation that it will not succeed in this mission too. One interview stated:

> If we look at the attempts made to localize other technologies, you will find that our performance was not quite good. Take automobile manufacturing as one example. There were ambitious efforts and we were optimistic, but it had failed. Maybe renewable energy is a different case because back then the government was not really into manufacturing and localization but now it is. Either way, localization of renewable energy technologies has to be supported by all means to take advantage of high potential Saudi Arabia has in this sector." (Interview no. 26– R&D stakeholders).

During conducting the interviews, a loophole that could possibly hinder the development of a solid base of solar technology local content was identified. When tendering a project, MOE requires all developers to maintain no less than 30% local content of the entire project's budget, without specifying what components are required to be local content. This allows investors and developers to fulfill this stipulated condition through infrastructure and construction content (instead of solar energy technologies), which easily accessible in the local market and does not effectively reflect on the local content targets set by NREP.

In summary, Saudi Arabia has a promising opportunity not only to develop a solid base of local content for solar energy but also to position itself as a solar technology hub for the MENA region. Although it is not clear yet how the 200 GW-solar projects would influence the development of the local content, it may significantly help accelerate the localization of this technology if appropriate local content policies are designed.

5.6.4 Summary of the Key Findings in the Analysis of Saudi Arabia Solar Energy System

With a particular focus on RSPSs, solar energy policies in Saudi Arabia have been analyzed exhaustively from various angles through the Sects. 5.6.1–5.6.3. Each subsection represents one layer of the *MLP* analytical tool encompassing several sub-questions that sought to capture the scene of the technology transition of RSPSs in Saudi Arabia by obtaining first-hand data. The key findings of the analysis can be summarized by each subsection as follows:

Landscape (macro-level)

It has become evident that the political buy-in into solar energy is the main driving force of the dissemination of this technology. While the top-down structure, weak bottom-up innovation, and limited involvement of the CSOs arguably remain as challenges, they may not necessarily act as formidable obstacles to the development of the solar energy industry in Saudi Arabia. On the contrary, the nature of a "globalizing monarchy" led by young, ambitious rulers may allow for a swift passing of new laws, policy amendments, and institutional reform to enhance the implementation of solar energy, which likely takes a longer time in other democratic countries (***SQ1***). This, however, does not mean the uncertainty or risk revolve around the solar energy projects are less, nor does it mean the goals set for solar energy are not subject to moderate change caused by changes in the political system (***SQ2***).

Moreover, the decision-making structure and the decision-making processes of the energy systems, especially for utility-scale projects, have undergone a significant improvement to enhance the alignment with the national Vision 2030. Nevertheless, the power-centralized structure and weak cross-cutting cooperation between the related government entities and low accountability are still apparent, presenting unneglectable challenges (***SQ3***). Further, Saudi Arabia has demonstrated its ability to develop its national renewable energy plans with not much assistant from transregional energy governance bodies. While some of these transregional organizations have worked with server Arab countries to identify and set their renewable energy strategies and objectives, their influence on the solar energy policies in Saudi Arabia is confined in issuing reports and recommendations (***SQ4***).

It has also been apparent that the economic necessity is the primary force for the development of the solar energy sector, and both the climate change agenda and the oil prices are "*secondary*" drivers for this development (***SQ5***, ***SQ6***). However, despite the potential benefits the solar energy ambitious plans can bring to the country by reintroducing it as a world peace advocate and eco-friendly country, no efforts seem to be made to link this distinctive development of solar energy to the country brand image, which indicates to an area of untapped potential (***SQ7***).

Regime (meso-level)

The national grid can take up 30% of renewable electricity before a further capacity extension needs to be implemented. The recently awarded project of installing 10 million smart meters by 2023 is expected to enhance the penetration of solar energy, however, it is questionable if that would be the case when the 200 GW project is fully implemented (***SQ8***). Saudi Arabia also bets on other forms of energy sources, such as nuclear energy, which is perceived as a complementary option. Despite the tremendous risk of holding nuclear reactors in an unstable region, such as the Middle

East, the motive for such an action might not be entirely economic, but also political (***SQ9***).

Additionally, despite the signs of failure in deploying RSPSs in Saudi Arabia related entities, such as the regulatory body, ECRA, and the service provider, SEC, did not exhibit any intention to lead new promotion campaigns for the RSPSs (***SQ10***). Further, the overwhelming majority of the interviewees underline the importance of indigenously developed policy models for the RSPSs based on a well-understanding of the lessons learned from the global experiences, rather than copying policies from other countries in which RSPSs have been successfully deployed (***SQ11***). With the basic regulations in place, and with the perception of a good cognition level of solar energy among individuals (***SQ13***), further efforts need to be made to ensure the economic viability of RSPSs against other subsidized conventional energy sources (***SQ12***). Other approaches can be taken to stimulate the adoption of RSPSs. For instance, although the government has been using religious materials (e.g., verses of the Qur'an and Hadiths[18]) in campaigns to enhance the rationalization of using the utility services such as water consumption, the results of this study revealed that there are no such established activities in the case of RSPSs, as confirmed by the stakeholders, some of whom were strongly in acceptance of using religion as a promoting tool (***SQ14***). Based on the findings in (***SQ1***, ***SQ7***, ***SQ14***), the hypothesis (H6) might be validated.

Further, while the recent efforts of implementing governance models started to results notable progress in terms of the utility-scale IPPs execution, neither identified goals nor long-term strategy is established to navigate the road of how RSPSs should be implemented in alignment with the national renewable energy plans (***SQ15***). This finding might support the validation of the hypothesis (H5). Moreover, it has become apparent that the social aspect is being considered with seriousness when energy subsidies are abolished, and no vision has been formed yet of whether new subsidies should be allocated to RSPSs (***SQ16***).

Niche (micro-level)

Additionally, it has been evident that there are no current endeavors to create a domestic knowledge hub for renewable energy technologies, nor there is an industry structure that stimulates the dissemination of technologies into mainstream society (***SQ17***). Further, among the technology choices Saudi Arabia can harness, more opinions were in favor of PV technology for utility-scale projects (***SQ18***). Finally, with the large capacity of solar energy expected to be installed in the near future, Saudi Arabia, as a country with an abundant stock of pure materials used in the manufacturing of PV panels, is prepared to develop a robust base of local content through obligations appropriately designed for project developers (***SQ19***) (Table 5.6).

[18]See Footnote No. 21.

Table 5.6 Key findings of Saudi Arabia's solar RSPSs policy analysis

Challenge	Description	Findings
1	A dominant top-down decision making and power-centralized structure	Regime actors' buy-in of a certain form of energy might be at the expense of others; bureaucratic hurdles; weak cross-cutting cooperation between the relative government entities
2	Command-and-control energy regulation versus market-driven energy regulation	Energy policies designed for the residential sector are in the form of governmental regulations rather than laws; special focus on public services-related buildings
3	Weak engagement of the non-governmental stakeholders in residential solar energy policy decisions	Limited influence of independent CSOs, transregional energy governance bodies and businesses, except for a few large-scale companies
4	Conflict of interests among pertinent stakeholders	Double subsidies; electricity market liberalization; decentralization; weak transparency
5	Absence of vision to navigate RSPSs implementation in harmony with the national utility-scale solar energy plans	Clear vision and strategies for the implementation of utility-scale solar projects, but as of yet, not for RSPSs
6	A loophole in the local content obligation	Successful bidders tend to take advantage of the unspecified local content obligation by fulfilling it thought construction component, instead of solar energy technologies
7	Adherence to the course of national plans	Low accountability of high-level decision-makers
8	Information provision on solar PV systems	Several failed attempts to create information portals and solar energy-related CSOs
9	The untapped potential of religion and country branding	Low willingness to use religion to promote the adoption of RSPSs; poor linking between solar energy dissemination and country branding

5.7 Policy Design Modeling of Transition Management

By combining the finding of the RSPSs policy analysis with the ideal type *TM* policy design the following situation is developed (Table 5.7).

While the *MLP* turned out to be an adequate model for an interdisciplinary analysis of the RSPSs energy system in Saudi Arabia, the applicability of the ideal type *TM* policy design and addressing the presented caveats remain subject to how the Saudi state will proceed with the matter of liberalizing the electricity market and the engagement of CSOs and non-business stakeholders.

Table 5.7 TM policy design, realpolitik caveats, and expected outcomes

Ideal type *TM* policy design	Realpolitik caveats	Expected outcomes
Step (1) Problem structuring in the transition arena and formulating a vision		
Sets up a multi-stakeholder energy advisory committee within REPDO dedicated to stimulating the adoption of RSPSs (including representatives of MOE, MLSD, SASO, K.A.CARE, KACST, ECRA, SEC, KAUST, and KAPSARC)	Applicable but not fully implemented	Potential scenarios for RSPS integration are developed, a long-term vision is formed and policies are codified
The advisory committee actively involves representatives of CSOs, solar energy SMBs and non-business stakeholders such as environmental and religious groups in periodic meetings with the advisory committee to assure their views are integrated into the polices design process	• Applicable, however not carried out, although similar practices of requesting public comments on national plans have been observed in other sectors • Restrictive legal frameworks impede the formation and significative operations of CSOs	• Re-assessment of the current small-scale solar PV regulations • Limitations of the energy system, strengths, and weakness of draft policies are openly discussed and assessed
The advisory committee shall identify necessary changes to the current small-scale solar PV regulations	Implementable, however, it may require a long time due to the energy market constraints	Alignment formation between relevant bodies
Step (2) Devising a concrete transition pathway and agenda		
The advisory committee shall set short, mid, and long-term goals of RSPSs integration of the country's energy mix	Applicable, however, the non-existence of defined planned capacity of RSPSs, and the absence of a clear vision of how both small-scale and utility-scale PV system capacity should be integrated may hinder developing defined goals	Defining clear goals will increase transparency, reduce investment uncertainty, and will help design a robust national energy plan
The advisory committee shall update the current small-scale solar PV regulations	Implementable, however, it may require a long time to achieve market liberalization, and reach a settlement of the conflict of interests between relevant bodies	Fairer energy regulations developed based on the lesson learned from the previous version that would encourage the adoption of RSPSs

(continued)

Table 5.7 (continued)

Ideal type *TM* policy design	Realpolitik caveats	Expected outcomes
The advisory committee shall add more specification to the local content obligation by which developers are compelled to include solar system components manufactured locally	Applicable, yet not exist	• Create a local demand base and enhance the localization of the solar energy system components • Boost the development of solar energy sector and contribute to job creation
To develop a national interactive platform under the supervision of the advisory committee, to provide information on solar PV system to the public	Applicable, yet not exist	Allowing individual households to develop a piece of knowledge in advance (about required upfront investment, required space, estimated electricity output, and potential revenue based on dwelling location) will assist them to have an intact decision-making capacity regarding the adoption of RSPSs
Reflecting on global experience, the advisory committee shall develop indigenous solar PV uptake schemes and promoting mechanisms including tax deduction, financial incentives, time-of-use rates, and soft loans (derived from electricity tariff correction)	Applicable, yet not exist (except for Net Metering mechanism)	• Political economy and citizen-state relation balance • Developing local solar energy industry and boosting the growth of solar energy SMBs • Easing the pressure on the national grid, especially on peak demand times
Step (3) Experimental phase		
Based on the revised policy document, the updated measures are applied to a small-scale area or city to examine their effectiveness and competence[a]	Applicable, especially to housing units equipped with smart meters	Precaution to prevent oversupply or grid fluctuation
Step (4) Monitor and evaluate the results		
Performance is monitored and results are evaluated and compared to the expected outcomes	Applicable if the measures are implemented in a small-scale area	Identifying opportunities for improvement or any system loopholes responsible for poor performance

(continued)

Table 5.7 (continued)

Ideal type *TM* policy design	Realpolitik caveats	Expected outcomes
Step (5) Adaptation		
By reflecting on the potential flaws of the previous step, the advisory committee shall develop the measures to mitigate any adverse effects	Applicable	• Redressing the system's failures and developing sustainable measures • Ensure to prevent the recurrence of the identified issues
Step (6) Reiterate from step (1)		
To reiterate the setps after the advisory committee confirms its validity	Applicable	Proceed in implementing the measures with a better base-line

[a]When the ideal type *TM* policy design is implemented as multiple iterations, the 2nd iteration of the *Experimental phase* can be implemented in a different area (s), instead of being implemented in the same area with the tiered installation of solar PV system capacity. The rationale behind this is that ECRA sets caps on solar PV installations to not exceed 15% of the distribution transformer capacity in any area. Further, the maximum aggregate capacity to be installed in the area of any electricity department would be up to 5 MW.

5.8 Recommendations

In this chapter, the author analyzed the renewable energy system in Saudi Arabia and potential factors shape its policies. Since Saudi Arabia shares many characteristics with other hydrocarbon-rich countries, especially, Arab countries, conducting a comparative analysis with studies conducted a similar context such as the study on UAE and Algeria that was by Kumetat (2012), could be of interest. Moreover, it would be worthwhile conducting an empirical study on the Saudi case, yet by testing a policy design model other than the *TM* approach. Doing so could help to address the critics towards the *TM* applicability to non-democratic power-cartelized countries. Moreover, it could be of interest to conduct an in-depth analysis of local and transregional solar developing companies from different sizes as it would help to gain more insights into the business environments condition, local content development, and government tender awarding competitiveness.

5.9 Limitations

As stated earlier, it is not the goal of the study to draft one national energy policy of RSPSs, as this involves law reforms and more technical work on the electricity market liberation to achieve the LCOE. This would also require technical work related to the PV systems' efficiency and other matters discussed in Sect. 5.6. This study, however, by using *MLP* aimed to shed light on the key governance features that need to be considered by presenting critical obstacles in their real-life settings. Further, the *TM* policy design model has helped to identify which governance structures represent realpolitik caveats that needed to be addressed drawing a picture of the critical conditions of success of RSPSs energy policy in Saudi Arabia. Having said that, the applicability of the proposed *TM* policy may not be as effective as it is when applied to a democratic open liberal government. Additionally, although this study interviewed several local developers, an in-depth analysis of the obstacles they confront was not fully explored as it is beyond the scope of this study. The study in this chapter limited itself to analyzing the Saudi case, without including a comparative analysis of a study done in different settings—which may be considered as one of the limitations.

5.10 Conclusion

In this chapter, the *MLP* governance model was used as an analytical tool to analyze the residential solar photovoltaics systems policies in Saudi Arabia. Through conducting face-to-face semi-structured interviews with stakeholders in the renewable energy sector, first-hand data was collected and analyzed, and conclusions were drawn. Based on the findings of this analysis, the energy policy system of the RSPSs can be characterized as a top-down and power-centralized decision-making structure coupled with weak cross-cutting cooperation between the related governmental entities.

Additionally, there is a high willingness to change this situation, yet, the absence of a unified reference point hinders the formulation of a long-term strategy that integrates the RSPSs capacity with the national renewable energy plans. Although the implementation of RSPSs can help balance out the citizen-state relation and supports the government's efforts in minimizing the concept of the "rentier state", the RSPSs adoption rates are less likely to increase under the current Net Metering mechanism offered against substantially subsidized electricity generated by conventional means. Besides the lack of effective incentivizing mechanisms, the weak involvement of the CSOs and the disregard of religion-based, and country branding-based promotion approaches have been found as areas of untapped potential. Based on the results of this study, a set of recommendations and guidelines were put forward to enhance the RSPSs dissemination in a suitable manner.

References

Aawsat [WWW Document]. (2016). The full transcript of Al-Arabiya's interview with the Saudi crown prince. Retrieved December 14, 2019, from https://www.shorturl.at/gnY06.

Al-Barrak, S. (2018, April 25). Personal communication with Shareef Al-Barrak, the Vice Director Technical Affairs of Electricity & Cogeneration Regulatory Authority ECRA, at the headquarters located in Riyadh.

Albert, A. (1995). *Chaos and society, frontiers in artificial intelligence and applications*. IOS Press.

Almasoud, A. H., & Gandayh, H. M. (2015). Future of solar energy in Saudi Arabia. *Journal of King Saud University-Engineering Sciences, 27,* 153–157. https://doi.org/10.1016/j.jksues.2014.03.007.

Alrashoud, K., & Tokimatsu, K. (2019). Factors influencing social perception of residential solar photovoltaic systems in Saudi Arabia. *Sustainability, 11,* 5259. https://doi.org/10.3390/su11195259.

Al-Suwaiyel. (2015). Science, technology and Innovation in Saudi Arabia, Global Competitiveness Forum [WWW Document]. Retrieved December 20, 2019, from https://slideplayer.com/slide/3809174/.

Al-Taher, M. (2018, May 14). Personal communication with Mohammed Y. Al-Taher, the Distribution Planning Department Manager of Saudi Electricity Company SEC, at the headquarters located in Riyadh.

Al-Turaiqi, A. (2008). *The political system of Saudi Arabia* (1st ed.). Ghainaa Publications.

Alwasatnews [WWW Document]. (2015). Retrieved November 4, 2019, from http://www.alwasatnews.com/news/988136.html.

Amen, S. A. (2017). Economic diversification policies of income sources in Kingdom of Saudi Arabia: Analytical reading. *Economic Horizons, 38,* 65–106. https://doi.org/10.12816/0043709

Atomic Scientists. (2019). Bulletin of the Atomic Scientists. Retrieved November 8, 2019, from https://thebulletin.org/2019/09/decoding-the-us-secret-authorizations-to-sell-nuclear-technology-to-saudi-arabia/.

Ayaz, K., Akbar, G., & Aslam, A. (2003). World wide fund for nature (1st ed.). Conservation and Islam, Islamabad.

Badran, A., Murad, S., Baydoun, E., & Daghir, N. (2017). *Water, energy & food sustainability in the Middle East*. The sustainability triangle, Adnan Badran, (1st ed.). Springer International Publishing.

Bahgat, G. (2006). Nuclear proliferation: The case of Saudi Arabia. *Middle East Journal, 60,* 421–443.

Balea, J. (2015). *Can the church shift the balance to renewable energy in the Philippines?* The Guardian.

Berkhout, F. G. H., Smith, A., & Stirlingh, A. (2004). Socio-technological regimes and transition contexts. *System Innovation and the Transition to Sustainability*, 48–75.

Binder, C. R., Hinkel, J., Bots, P. W. G., & Pahl-Wostl, C. (2013). Comparison of frameworks for analyzing social-ecological systems. *Ecology and Society*, 18.

Binder, R. C., Mühlemeier, S., & Wyss, R. (2017). An indicator-based approach for analyzing the resilience of transitions for energy regions. Part I: Theoretical and conceptual considerations. *Energies*, 10. https://doi.org/10.3390/en10010036

Blazquez, J., Hunt, L., & Manzano, B. (2017). Oil subsidies and renewable energy in Saudi Arabia: A general equilibrium approach. *The Energy Journal, 38,* 29–45.

Burck, J., Marten, F., & Bals, C. (2015). *The climate change performance index results 2016.* Germanwatch

Casey, J. (2019). Road to nowhere: Great solar disappointments. Power Technology, Energy News and Market Analysis. Retrieved November 7, 2019, from https://www.power-technology.com/features/solar-failed-projects-struggling/.

Electricity & Cogeneration Regulatory Authority ECRA (2017). Small-Scale Solar PV Systems Regulations.

Elizabeth, S., & Gordon, W. (2007). CAUTION! Transitions ahead: Politics, practice, and sustainable transition management. *Environment and Planning, 39,* 763–770. https://doi.org/10.1068/a39310. Elizabeth Shove, Gordon Walker, 2007.

Erhun, K. (2001). Islam and environmental conservation. *Environmental Conservation, 28,* 1–9. https://doi.org/10.1017/S0376892901000017.

FCI [WWW Document]. (2019). FutureBrand. Retrieved November 6, 2019, from https://www.futurebrand.com/futurebrand-country-index.

Fiscal Balance Program [WWW Document]. (2017). Retrieved November 13, 2019, from https://vision2030.gov.sa/en/programs/FBP.

Foxton, T. J., Hammond, G., & Pearson, P. J. (2008). Transition pathways for a low carbon energy system in the UK: Assessing the compatibility of large-scale and small-scale options. In *Presented at the 7th BIEE academic conference.*

Frey, J., & Oishi, S. (1995). How to conduct interviews by telephone and in person. London: Sage Publications Ltd.

GAS [WWW Document]. (2019a). General Authority for Statistics. Retrieved November 9, 2019, from https://www.stats.gov.sa/en/indicators/224.

GAS. (2019b). 1.60% of households using solar energy at residence in 2019 [WWW Document]. General Authority of Statistics. Retrieved December 22, 2019, from https://www.stats.gov.sa/ar/1060.

Geels, F. W. (2002). Technological transitions as evolutionary reconfiguration processes: A multi-level perspective and a case-study. *Research Policy, 20–31*, 1257–1274. https://doi.org/10.1016/S0048-7333(02)00062-8

Geels, F. W. (2004). From sectoral systems of innovation to socio-technical systems: Insights about dynamics and change from sociology and institutional theory. *Research Policy, 33*, 897–920. https://doi.org/10.1016/j.respol.2004.01.015.

Geels, F. W. (2005). *Technological transitions and system innovations: A Co-evolutionary and Socio-Technical Analysis*. Edward Elgar Publishing Ltd.

Geels, F. W. (2011). The multi-level perspective on sustainability transitions: Responses to seven criticisms. *Environmental Innovation and Societal Transitions, 1*, 24–40. https://doi.org/10.1016/j.eist.2011.02.002.

Geels, F. W. (2014). Regime resistance against low-carbon transitions: Introducing politics and power into the multi-level perspective. *Theory, Culture & Society, 31*, 21–40. https://doi.org/10.1177/0263276414531627.

Geels, F. W., & Schot, J. (2007). Typology of sociotechnical transition pathways. *Research Policy, 36*, 399–417. https://doi.org/10.1016/j.respol.2007.01.003.

Gonand, F., Hasanov, F., & Hunt, L. (2019). Estimating the impact of energy price reform on Saudi Arabian intergenerational welfare using the MEGIR-SA Model. *The Energy Journal, 40*, 55–77. https://doi.org/10.5547/01956574.40.3.fgon.

Grin, J., Rotmans, J., & Schot, J. (2011). On patterns and agency in transition dynamics: Some key insights from the KSI programme. *Environmental Innovation and Societal Transitions, 1*, 76–81. https://doi.org/10.1016/j.eist.2011.04.008.

Häfele, W., & Sassin, W. (1977). The global energy system. *Annual Review of Energy, 2*, 1–30. https://doi.org/10.1146/annurev.eg.02.110177.000245.

Henry, C. M., & Springborg, R. (2001). *Globalization and the politics of development in the Middle East, the contemporary Middle East*. Cambridge: Cambridge University Press. https://doi.org/10.1017/CBO9780511807688.

Hertog, S. (2010). *Princes, brokers, and bureaucrats: Oil and the state in Saudi Arabia* (1st ed.). Cornell University Press.

Jochen, M., Rob, R., & Bernhard, T. (2012). Sustainability transitions: An emerging field of research and its prospects-ScienceDirect, *41*, 955–967. https://doi.org/10.1016/j.respol.2012.02.013

John, G., Jan, R., & Johan, S. (2010). Transitions to sustainable development: New directions in the study of long …. New York, USA: Google Books, Routledge.

K.A.CARE [WWW Document] (2017). Retrieved November 7, 2019, from https://www.energy.gov.sa/en/projects/Pages/atomic.aspx.

Kamrava, M., Nonneman, G., Nosova, A., & Valeri, M. (2016). Ruling families and business elites in the Gulf Monarchies: Ever closer?

Kemp, R. (2010). The Dutch energy transition approach. *International Economics and Economic Policy, 7*, 291–316. https://doi.org/10.1007/s10368-010-0163-y.

Kemp, R., Loorbach, D., & Rotmans, J. (2007a). Transition management as a model for managing processes of co-evolution towards sustainable development. *International Journal of Sustainable Development and World Ecology, 14,* 78–91. https://doi.org/10.1080/13504500709469709.

Kemp, R., Rotmans, J., & Loorbach, D. (2007b). Assessing the Dutch energy transition policy: How does it deal with dilemmas of managing transitions? *Journal of Environmental Policy & Planning, 9,* 315–331. https://doi.org/10.1080/15239080701622816.

Kern, F. (2010). *The politics of governing 'system innovations' towards sustainable electricity systems (doctoral).* University of Sussex.

Kiel, L. D., & Elliott, E. W. (1997). *Chaos theory in the social sciences: Foundations and applications.* University of Michigan Press.

Kumetat, D. (2012). Managing the transition: An analysis of renewable energy policies in resource-rich Arab states with a comparative focus on the United Arab Emirates and Algeria. The London School of Economics and Political Science, London.

Loorbach, D. (2007). Transition management: New mode of governance for sustainable development.

Loorbach, D. (2010). Transition management for sustainable development: A prescriptive, complexity-based governance framework. *Governance, 23,* 161–183. https://doi.org/10.1111/j.1468-0491.2009.01471.x.

Manu, B. (2016). Retrieved November 10, 2019, from https://qz.com/india/779196/narendra-modi-turns-to-the-sun-india-is-going-to-use-godmen-to-spread-the-use-of-solar-energy/.

Meadowcroft, J. (2005). Environmental political economy, technological transitions and the state. *New Political Economy, 10,* 479–498. https://doi.org/10.1080/13563460500344419.

Meadowcroft, J. (2011). Engaging with the politics of sustainability transitions. *Environmental Innovation and Societal Transitions, 1,* 70–75. https://doi.org/10.1016/j.eist.2011.02.003.

Mosly, I., & Makki, A. (2018). Current status and willingness to adopt renewable energy technologies in Saudi Arabia. *Sustainability, 10,* 4269. https://doi.org/10.3390/su10114269.

Musselwhite, K., Cuff, L., McGregor, L., & King, K. M. (2007). The telephone interview is an effective method of data collection in clinical nursing research: A discussion paper. *International Journal of Nursing Studies, 44,* 1064–1070. https://doi.org/10.1016/j.ijnurstu.2006.05.014.

National Transformation Program [WWW Document]. (2016). Retrieved November 4, 2019, from https://vision2030.gov.sa/en/programs/NTP.

Normann, H. E. (2015). The role of politics in sustainable transitions: The rise and decline of offshore wind in Norway. *Environmental Innovation and Societal Transitions, 15,* 180–193. https://doi.org/10.1016/j.eist.2014.11.002.

OEC [WWW Document]. (2019). Retrieved November 5, 2019, from https://oec.world/en/profile/country/sau/.

Osunmuyiwa, O., Biermann, F., & Kalfagianni, A. (2018). Applying the multi-level perspective on socio-technical transitions to rentier states: The case of renewable energy transitions in Nigeria. *Journal of Environmental Policy & Planning, 20,* 143–156. https://doi.org/10.1080/1523908X.2017.1343134.

pv magazine [WWW Document]. (2019). pv magazine International. Retrieved November 7, 2019, from https://www.pv-magazine.com/2019/01/03/saudi-arabia-plans-to-set-new-long-term-re-targets-and-to-resume-tenders/.

qafilah, Scientific research indicators in the Kingdom. (2016). Retrieved December 20, 2019, from http://shorturl.at/afhnD.

Qatrah, National Water Company (2019). [WWW Document]. https://www.qatrah.com/pledge (accessed 12.13.19).

Renewable Energy Project Development Office REPDO (2017). Ministry of Energy, Industry and Mineral Resources [WWW Document]. https://www.powersaudiarabia.com.sa/web/index.html (accessed 8.21.19).

Renzetti, C., & Lee, R. (1993). *Researching sensitive topics, CRVAW faculty book gallery* (14th ed.). Newbury Park, California: Sage Publications.

Reuters, Saudi prince unveils sweeping plans to end "addiction" to oil [WWW Document]. (2016). Retrieved November 1, 2019, from https://www.reuters.com/article/us-saudi-economy/saudi-prince-unveils-sweeping-plans-to-end-addiction-to-oil-idUSKCN0XM1CD.

Rip, A., Kemp, R. P. M., & Kemp, R. (1998). *Technological change. Human choice and climate change* (Vol. II, pp. 327–399). Resources and Technology.

Rolffs, P., Ockwell, D., & Byrne, R. (2015). Beyond technology and finance: Pay-as-you-go sustainable energy access and theories of social change. *Environment and Planning A, 47,* 2609–2627. https://doi.org/10.1177/0308518X15615368.

Rotmans, J., & Loorbach, D. (2008). Transition management: Reflexive governance of societal complexity through searching, learning and experimenting. *International Journal of Sustainable Development and World Ecology.*

Rotmans, J., Kemp, R., & Asselt, M. (2001). More evolution than revolution: Transition management in public policy. *Foresight, 3,* 15–31. https://doi.org/10.1108/14636680110803003

Saudi Arabia may become oil importer by 2030: Citigroup, Financial Post [WWW Document]. (2012). Retrieved November 1, 2019, from https://business.financialpost.com/commodities/energy/saudi-arabia-may-become-oil-importer-by-2030-citigroup.

Saudi Arabia Solar Industry Association (SASIA)-Home page [WWW Document]. (2011). Retrieved December 10, 2019, from https://www.facebook.com/Saudi-Arabia-Solar-Industry-Association-SASIA-165029500281965/.

Saudi Electricity Company SEC (2019). Smart meters project [WWW Document]. https://www.se.com.sa/en-us/pages/newsdetails.aspx?NId=954 (accessed 12.20.19).

Schäpke, N. (2018). *Linking transitions to sustainability: Individual agency, normativity and transdisciplinary collaborations in transition management.* ZBW-Leibniz Information Centre for Economics.

Sheng, C., Cao, Y., & Xue, B. (2018). Residential energy sustainability in China and Germany: The impact of national energy policy system. *Sustainability, 10,* 4535. https://doi.org/10.3390/su10124535.

Tlili, I. (2015). Renewable energy in Saudi Arabia: Current status and future potentials. *Environment, Development and Sustainability, 17,* 859–886. https://doi.org/10.1007/s10668-014-9579-9.

Two countries that prove nation branding works, Jose Torres, Bloom Consulting, City Nation Place [WWW Document]. (2020). Retrieved January 18, 2019, from https://www.citynationplace.com/two-countries-that-prove-nation-branding-works.

Ulsrud, K., Winther, T., Palit, D., Rohracher, H., & Sandgren, J. (2011). The solar transitions research on solar mini-grids in India: Learning from local cases of innovative socio-technical systems. *Energy for Sustainable Development, 15,* 293–303. https://doi.org/10.1016/j.esd.2011.06.004. Special issue on off-grid electrification in developing countries.

Verbong, G., & Geels, F. (2007). The ongoing energy transition: Lessons from a socio-technical, multi-level analysis of the Dutch electricity system (1960–2004). *Energy Policy, 35,* 1025–1037. https://doi.org/10.1016/j.enpol.2006.02.010.

Wimmer, A., & Kössler, R. (2005). Understanding change: Models, methodologies and metaphors. Springer.

Yin, R. K. (2009). *Case study research: Design and methods.* SAGE Publications, Inc.

Young, O. R., Lambin, E. F., Alcock, F., Haberl, H., Karlsson, S. I., McConnell, W. J., et al. (2006). A portfolio approach to analyzing complex human-environment interactions: Institutions and land change. *Ecology and Society, 11,* 31. https://doi.org/10.5751/ES-01799-110231.

Chapter 6
Discussion and Conclusion

In this chapter, the results of the previous chapters are summarized under each research question. Next, an evaluation of the study hypotheses is present. Furthermore, the overall conclusion and remarks are highlighted. Moreover, the key contributions and limitations of the study are also highlighted, and recommendations for future research are suggested.

6.1 Results of the Study

This book aims at advancing knowledge regarding the technology transition of RSPSs through investigating the social perception towards it and analyzing the factors that shape its policies in Saudi Arabia. It also puts forward a set of recommendations and guidelines to enhance the dissemination of RSPSs at the households' level. In this section, the results will be presented as answers to the questions posed by the research.

(RQ1). How individuals in Saudi Arabia perceive energy options?

This book has attempted to answer this question by investigating the perception of the individuals targeted in this study from various aspects. The respondents showed a high level of cognition regarding environmental and energy matters. Although their opinions on the longevity of crude oil reserves in Saudi Arabia have varied, a general preferential trend for changing from oil dependence to other energy alternatives was observed. Moreover, the majority of the respondents acknowledged that Saudi Arabia suffers from air pollution, agreeing that the leading cause of global warming was the result of fossil fuel emissions caused by coal, petrol, and diesel, an issue that they, by a large margin, recommended addressing by expanding the use of renewable energy.

K. A. Alrashoud, *Managing Technology Transition in Saudi Arabia*,
https://doi.org/10.1007/978-981-15-5803-0_6

(RQ2). To what extent do individuals in Saudi Arabia accept the notion of installing a residential solar PV system?

The results revealed that the overwhelming majority are associated with RSPSs with a positive image. The general trend in the data shows good awareness of solar energy. Most respondents agreed on the continuity of solar energy-based electricity supply, and the role of solar energy in job creation. On the other hand, they disagree that solar energy could negatively impact the environment and or local communities. Similarly, on the personal level, when the respondents were asked whether they accept or reject the notion of installing an RSPS, the overwhelming majority of them were in acceptance to adopt this technology.

(RQ3). What are the most critical factors influencing the individuals' decision to accept or reject the adoption of a residential solar PV system in Saudi Arabia?

Based on the results, the innovation attribute of *relative advantage* was the most critical determining factor among the other attributes. Additionally, the most significant variable among all the four attributes was the effect on *global warming*. The second most significant variable was *unlimited power*, and the third was *revenue*. On the other hand, the *installation cost* of the *compatibility* attribute was the leading constraint that can drive participants to reject RSPS installation.

(RQ4). How should the energy policies of the residential solar PV systems and the transition towards it in Saudi Arabia be understood?

Although a decade ago, solar energy used to be perceived as a threat to the economy's backbone of the country—fossil fuel, it is currently widely supported by top-level political rulers. Since 2016, Saudi Arabia has exerted substantial efforts to enhance the coordination between energy-related government entities and to streamline the decision-making structure of the energy systems. While these efforts are considered steps in the right direction, they do not seem to be going beyond the implementation of the utility-scale projects to reflect on the RSPSs deployment. Nevertheless, in a "globalizing monarchy" led by young, ambitious rulers, these challenges may not necessarily act as daunting challenges (as it used to be), since those leaders are exhibiting a political buy-in for solar energy. Given the enormous changes made in the past three years to modernize Saudi Arabia, the nature of a centralized-power structure may, on the contrary, boost the implementation of this technology as involves a small number of decision-makers, which thus, could allow for a swift enacting of new regulations or policy amendments.

A collaborative work between the pertinent entities has resulted in issuing basic regulations that manage the relationship between the end-users and the electricity service provider. Nevertheless, essential measures such as introducing incentive schemes, reliable RSPSs-related information references, public awareness, and promotion campaigns are still missing, which resulted in a significantly limited increase in the RSPSs adoption rates. This dilemma is arguably caused by the absence of a clear vision and that could have been elucidated how the RSPSs should be implemented in line with the national renewable energy plans. While this technology is intimately related to individuals, the involvement of the non-business groups in policy formulation and promotion activities remains limited, a fact that arguably does not seem to change any time soon. Further, Saudi Arabia has demonstrated its capability to develop its national renewable energy plans, with a limited influence of transregional energy bodies confined in issuing reports and recommendations.

Moreover, while the heavy electricity subsidies have been gradually eliminated, further efforts need to be made to ensure the economic viability of RSPSs. This area has great implications for the socio-political system in Saudi Arabia which seemed, during interviewing the government stakeholders, an area that is considered very carefully. Additionally, it has been evident that there are no current endeavors to create a domestic knowledge hub for renewable energy technologies, nor there is an industry structure that stimulates the dissemination of technologies into mainstream society. As a country with an abundant stock of pure materials used in the manufacturing of PV panels, boosted by the large capacity of solar energy expected to be installed in the next few years, Saudi Arabia is prepared to develop a solid base of local content through obligations appropriately designed for project developers.

With the dynamic changes the energy landscape in Saudi Arabia is witnessing, it is hard to predict the future trajectory of the RSPSs. However, if the obstacles impeding the deployment of the RSPSs are properly addressed, RSPS can play a critical role to balance out the relation between the government and its citizens and assists in minimizing the concept of the "rentier state."

6.2 Evaluation of Research Hypotheses

Based on the results, testing the validity of each of the postulated research hypotheses are presented as follow:

(H1) There is a specific preference among the Saudi population against using crude oil.

Although the extensive revenues from fossil fuel exports have been guaranteeing the Saudi's residents abroad range of welfare and prosperity, they think that there are available alternatives to oil (84.1%). Further, the majority of them (71.9%) do not prefer oil to continue as the primary energy source (95% CI: 69.6% to 74.2%). While (75%), strongly agreed or agreed that the leading cause of global warming was the result of fossil fuel emissions. Most of them (65.1%) see the best approach to address this issue is to expand the use of renewable energy, on the one hand, on the other hand (6%), (7%), and (11%) believe that the "expansion of nuclear power," "reduction of electricity consumption," and "investment in research and development" was the appropriate approach, respectively. Based on these results, we reject the null hypothesis ($H_0 1$) that states, "There is no specific preference of the

Saudi population in the source of energy" and accept the alternative hypothesis (H1) of specific preference among the Saudi population against using crude oil.

(H2) The majority of the Saudi population is willing to install RSPS if they would receive subsidies equivalent to 40% of the total cost.

The results of investigating the individuals' willingness to adopt an RSPSs were encouraging. While (79.7%) of the respondents (95% CI: 77.7% to 81.7%) were willing to install an RSPS if they would receive 40% back in financial subsidies, approximately 20.3% (95% CI: 18.3% to 22.3%) rejected this offer. The acceptors were then asked about their motivations if they would only receive 20% financial subsidies instead of 40%. The majority of the 40% acceptors (69.3%, 95% CI: 66.7% to 71.9%) accepted the suggested 20% financial subsidy, while 30.7% did not see enough benefit in purchasing an RSPS if they would only receive 20% back in financial subsidies. On the other hand, respondents who rejected the 40% financial subsidy were offered an additional incentive of a 60% subsidy, but even then, only one-quarter of them accepted this offer (25.7%, 95% CI: 20.8% to 30.6%), and the majority of them (74.3%, 95% CI: 69.4% to 79.2%) rejected the offer.

This does not consistent with the null hypothesis ($H_0$2) which postulate "The Saudi population's willingness to adopt an RSPS does not change whether or not they receive subsidies equivalent to 40% of the upfront cost." and thus, we accept the alternative hypothesis (H2) of "The majority of the Saudi population is willing to install RSPS if they would receive subsidies equivalent to 40% of the total cost."

(H3) The majority of the Saudi population accepts the notion of installing a residential solar PV system in their homes.

The results of the analysis showed that the overwhelming majority of the respondents (97.46%) accept the notion of installing an RSPS in their residence (95% CI: 96.7% to 98.3%). Therefore, as the confidence interval confirms, the hypothesis (H3) might be validated.

(H4) The most significant deciding factor in accepting residential solar PV systems in Saudi Arabia is the *revenue* obtained by reducing electricity consumption and selling excess electricity to the service provider.

The association between attributes and RSPS acceptance shows that the *relative advantage* was the most critical determining factor since the difference in the median score between acceptors and rejectors was 9 points. Moreover, the most significant variable among all attributes was the effect on *global warming*, where the mean among the acceptors was 1.69 points more than that among the rejectors. The second most significant variable was *unlimited power*, and the third was *revenue. Income statistics monitoring*, *installation space*, and *land use* had the highest mean difference among *observability*, *complexity*, and *compatibility* variables, respectively. These results are not consistent with the postulated hypothesis (H4).

(H5) There are no identified goals of RSPS planned capacity or how it should be implemented in alignment with the national renewable energy plans.

The planned capacity of the IPP projects for solar PV has been revised and increased from (5.9 GW) to (20 GW) by 2023 and (40 GW) by 2030. This amendment was made early in 2019, which came one year after the announcement of the 200 GW solar project between Saudi Arabia and the SoftBank Group, while the delay in executing this sizable project has been fueling speculation that the project will be terminated. The RSPSs turned out to be swinging between these projects with a complete absence of identified goals for RSPSs installation capacity or how this installation capacity should be implemented in harmony with the national renewable energy plans. Thus, the hypothesis (H5) might be considered as validated.

(H6) The participation of the independent Civil Society Organizations in the formulation of the RSPSs policies is significantly limited, with a neglect of the soft-power elements, such as religion and country branding as potential promotion tools.

The results highlighted that the restrictive environment poses an obstacle for solar energy-related CSOs towards a meaningful contribution to the formulation of policies that directly affect the local developer and individuals. In the past three years, attempts have been made to establish local independent CSOs aiming to streamline the efforts between the SMBs and serve as information provision platform of the RSPSs components and job vacancies in this sector, however, these aims have yet to materialize. Given the fact that the contribution of the non-profit sector to GPD is estimated at 0.2% and planned to increase to merely 0.6% by 2020, it is hard to expect an effective contribution of the CSOs to the solar energy sector.

Regarding employing religion to promote the adoption of RSPSs, although several stakeholders were strongly in favor of using the religion as a promoting tool, it has been clear that there is no such approach being utilized to encourage the Saudi residences to adopt RSPSs. Similarly, despite the potential indirect benefits that the expected Giga-solar projects can bring to the country, by reintroducing it as a world peace advocate and eco-friendly country, such potential was found to be effectively linked with the efforts that aim to improve the country brand image. All these findings might assist in the validation of the hypothesis (H6).

6.3 Overall Conclusion

The objective of this research was to contribute to the academic discussion around the RSPSs in Saudi Arabia. This research examined various aspects of the social perception in relation to the adoption of RSPSs, a critical area of renewable energy acceptance that has rarely been investigated in Saudi Arabia. The study also analyzed the current technology transition of RSPSs and the factors shaping its policies by exploring the broader energy systems, helping to capture wide-ranging implications of the status quo of the RSPSs' development in Saudi Arabia. Further, the study also proposed policy recommendations based on the TM governance models. Therefore, this study could also contribute to the development of policy and regulation, allowing the consideration of a wider range of citizen-state perspectives with a potential reformulation of the RSPS policies.

The quantitative survey conducted among householders revealed that the overwhelming proportion of the respondents associate RSPSs with a significantly positive image, where the majority of them had sound views of global warming and the necessity to escape from fossil fuel as the main energy source of the country. Long-term nonfinancial benefits factors under the *relative advantage* innovation attribute such as *environment protection* or the advantage of receiving *unlimited power* have a higher correlation with acceptance RSPSs. Further, the majority of respondents (79.7%) were open to purchasing an RSPS if they could receive subsidies equivalent to 40% of the upfront cost. Positive attitudes toward RSPSs may act as a driving force but as of yet, it is not enough for adoption without introducing incentive schemes.

The results also revealed that the top-down structure, weak cross-cutting cooperation between the related government entities, and weak bottom-up innovation arguably remain challenges, yet, they may not necessarily act as formidable obstacles to the development of the solar energy industry. This particularly rings true in a "globalizing monarchy" such as Saudi Arabia, where these technologies turned out to be reinforced by a political buy-in. Further, the government's recent efforts to improve governance have reflected positively on the implementation of the utility-scale independent power producers projects. However, the absence of a long-term strategy and specific goals to navigate the road of how RSPSs should be implemented in alignment with the national renewable energy plan, are missing.

By taking advantage of the individuals' overwhelming positive perception towards RSPSs and by utilizing the country's financial power to disseminate this technology, the Saudi government can address the extremely wasteful consumption habits, and balance out the citizen-state relation to supports its underway efforts in minimizing the concept of the rentier state.

6.4 Key Contributions of the Research

Based on various modes of data collection and analysis conducted, this book constructed new insights into the analysis of technology transition in Saudi Arabia. The book looked at the RSPSs by addressing the multiple factors influencing the deployment of this technology within the incumbent energy regime. The study framework was developed to reflect the interactions between the social actors, technology, and institutions within a socio-technical system to address the gaps in the literature. The book contributions are abridged below.

Using solar energy as an alternative source of energy is a highly preferable option among the respondents, with little variation in acceptance or rejection rates among all the five targeted areas of the survey (93.2–98.6%). This implies that unlike other countries with autonomous areas (Such as, The United States or India) that employ region-specific energy policies, unified diffusion policies could be applicable to all regions across Saudi Arabia regardless of its vast geographical areas and variance of its climatic conditions. Further, out of the study sample, (79.7%) were open to purchasing an RSPS if they could receive subsidies equivalent to 40% of the upfront cost, regardless of the actual RSPSs low adoption rates (1.6%). This indicates a gap that can be addressed by re-designing the currently introduced mechanism of Net Metering. RSPSs adoption can be boosted through offering an upfront subsidy and by taking advantage of the supportive perception of protecting the environment and positive association with solar energy.

A good cognition level in relation to solar energy, where the majority (64–83%) of respondents are aware of the benefits, and 97% are associated with RSPSs with a positive image. The results also show high statistical significance for the differences between RSPS acceptors and rejecters in all innovation attributes ($p < 0.001$). Among the four attributes, the *relative advantage* had the highest average mean difference at (1.53). Based on these observations, key policies to spread RSPSs should include developing an interactive information platform to assist individuals to make sound purchasing decisions.

There is a need to establish an advisory committee that includes all the pertinent entities to develop a vision and identified goals of how the RSPSs should be implemented in line with the other national renewable energy plans (Landscape). National branding and religious approaches are areas of untapped potentials that can be utilized to promote the adoption of RSPSs in Saudi Arabia (Regime). More specification of the local content obligation is needed where solar energy systems-related components should be required as a part of the technology localization obligation (Niches).

6.4.1 Framework Applicability

This book argued that RSPSs are embedded in its social and institutional context and thus, a successful deployment requires changes in social actors' cognitions and the regime vested interests. Throughout the discussion of this book, light has been shed on the public perception of RSPSs, and the key material dimensions of solar energy policies in Saudi Arabia. To address the former, the *DOI* theory was adopted to explain how and why RSPSs would gain momentum and spreads through the population of Saudi Arabia. Being used in a wide range of fields, including agriculture products, e-services, medical treatment procedures, this theory turned out to be also adequately explaining how a population of an oil-rich country such as Saudi Arabia perceives RSPSs as a newly introduced technology.

Similarly, for analyzing the key material dimensions of solar energy policies in Saudi Arabia the *MLP* prominent transition framework was adopted. While the majority of the *MLP* applications took place in a context that reflects open, liberal governments, the author, by building on previous studies, has utilized the *MLP*'s great deal of flexibility to apply it to the Saudi case—the non-democratic country with a dominant top-down decision making and power-centralized structure. By including different explanatory elements in various dimensions, the adopted model of *MLP* sufficiently provided a comprehensive depiction of the broader energy systems, interaction between its actors, and potential factors shaping its policies.

6.4.2 Empirical Contribution

The key empirical contribution that this book has made is in the understanding of the current development of the transition to RSPSs in Saudi Arabia. It was argued that extant literature primarily focuses on technical aspects of the micro-scale solar PV systems. Since the development of solar energy becomes an increasingly imperative component of the Saudi Arabia energy mix, it is argued that by considering social actors' perception of RSPSs and making its policies a central focus, the book allows and provides novel insights into the current technology transition towards RSPSs.

At the household level, a distinct empirical contribution is made by conducting an in-depth large-scale investigation on the Saudi residence perception of RSPSs across the five main areas in Saudi Arabia. The analysis of Chap. 3 aimed to explore the public's views in Saudi Arabia on energy options through a quantitative approach. By using a cross-sectional survey, a particular focus was placed on five aspects including basic demographic characteristics, residential status, energy consumption, environmental and energy cognition, and willingness to adopt RSPSs. Fresh insights were presented on how the Saudi residence perceive the country's main source of power, as well as the main source of national income. An investigation was conducted on how the status of their residential units and the behavior of electricity consumption are associated with their preferences for energy options. The environmental and energy cognition were also assessed by investigating the individuals' level of knowledge on pollution issues, global warming, and potential solutions to address them. Additionally, questions of the individuals' willingness to adopt RSPSs were given as they provide important insights on how incentive mechanisms should be designed and offered. Answers to these questions contribute to understanding how willing the Saudi residents are to shift to RSPSs. Since this questionnaire was deployed after the recent government's announcement of allowing households to integrate RSPSs to their dwellings, the results of this investigation are arguably providing coherent up-to-date insights into this area.

On a deeper level, Chap. 4 investigates the precise factors that could potentially encourage or hinder individuals from adopting RSPSs. Based on an empirical investigation guided by the *DOI* theory, an in-depth analysis was conducting exploring a wide range of drivers and barriers that motivate or hamper, from the household perspective, the RSPSs diffusion. Empirical data were gained analyzed providing novel insights from approaches that have rarely been conducted in the literature in the Saudi context.

Another empirical contribution of this book is, collecting and analyzing first-hand data from the stakeholders of the renewable energy sector (public sector, private sector, and R&D sector) (Chap. 5). Conducting face-to-face interviews has significantly contributed to a deeper understanding of the solar energy landscape in general, in the RSPSs' in particular, providing valuable insights for the ongoing processes of the development of the RSPSs and associated obstacles and opportunities. The opaque nature of Saudi Arabia has been an obstacle for external researchers to conduct their work, especially researches that involve politically sensitive topics. However, in this study, the author has successfully met with a sufficient number of informed stakeholders who provided saturated data that was used, in addition to the empirical data gained in Chaps. 3 and 4, to draw a comprehensive picture of the RSPSs transition in Saudi Arabia.

Besides the novel insights this study revealed, resulting in benefits to the local community, the generalization of these findings to other oil-exporting countries are guided by similarities in context, specifically low-priced electricity tariffs and per capita income. It is hoped that this endeavor will pave the way for further studies and help citizens and policymakers in addressing the challenges hampering the dissemination of RSPSs.

6.5 Limitations and Avenues for Future Research

The empirical results presented in this book should be considered in light of some potential limitations that need to be addressed to enhance future work.

It is important to note that the obstacles to the implementation of renewables differ across different technologies and, thus, further work is required to define, recognize, and mitigate the barriers by taking into account the nature of each technology. This also rings true for solar energy as this study was dedicated to the case of RSPSs, there were some unavoidable overlapping areas between the analysis of solar energy policies and the RSPSs policies. Thus, it should not be considered that the analysis conducted in this study fully explored the policies of solar energy in Saudi Arabia.

With regard to the cross-sectional survey conducted among householders, since it was conducted online, it was not applicable to those who do not have access to the internet. Further, the online approach was reflected in various demographic characteristics of the respondents, even though controlling for these demographics indicated no significant variations in their subcategories. Further, the types of questions used in the questionnaire were closed-ended, Likert type scale questions and a dichotomous question. While these types of questions provide precise and clear answers, they do not offer a specific explanation of why a certain answer was chosen among other options. Thus, an open-ended, face-to-face approach would provide deep insight by allowing respondents to define central issues and explain the rationale behind their answers.

Moreover, Missing data occurred for two questions. However, the analysis of the other variables indicated that these were Missing Completely at Random (MCAR). This claim is supported by the fact that they were not significantly associated with any other variable. The MCAR notation generally does not cause significant changes or bias in the results of the survey questions. These missing data could have resulted from an unknown system error in the online survey form, supporting their classification as MCAR.

Moreover, although the study in Chap. 5 interviewed a number of local developers, barriers hampering business from their perspectives were not fully explored. Similarly, the financial institutions which play an important role in assessing the risk of each renewable project and offering loan programs were not included within the scope. A study shedding light on these areas would assist to identify and overcome obstacles spreading throughout the entire solar PV value chain.

Since Saud Arabia is in the early stages of implementing RSPSs, further work needs to be done to understand the process of the development of this technology. Analysis of the households' consumption behavior before and after the adoption of RSPSs could an interesting area to investigate. Such analysis could help understanding not only if their consumption behavior has changed, but also to what extent it might have changed. While similar studies have been conducted in the European context (McKenna and Thomson 2014; Moezzi et al. 2017), it could be of interest to investigate how the power consumption behavior of RSPSs adopters in the hydrocarbon-rich countries might have changed.

Another future research direction is investigating the functionality between the relevant government agencies based on an innovation-oriented theoretical framework that may help elucidate how the involved parties should work jointly to enhance the inducement mechanisms and reduce the blocking mechanisms that have a great impact on the diffusion of RSPSs in Saudi Arabia. Furthermore, an in-depth investigation of the willingness to pay could be an interesting point, providing more insight into the implication of a Saudi residence adopting an RSPS. Another aspect that could be of interest is the electricity consumption behavior among households who already installed RSPSs; a study could be conducted to analyze how their electricity consumption might have changed after installing the solar PV system.

6.6 Recommendations

Based on the findings of this research and reflecting on the proposed *TM* police design model in Chap. 5, some recommendations can be derived for the policymakers in Saudi Arabia to facilitate the adoption of RSPSs.

- There is a need to establish a multi-stakeholder energy advisory committee within REPDO (including representatives of MOE, MLSD, SASO, K.A.CARE, KACST, ECRA, SEC, KAUST, and KAPSARC), where this committee fully dedicated to stimulating the adoption of RSPSs.
- The advisory committee shall formulate a vision that clearly articulates how should RSPSs be implemented in harmony with the other national renewable energy plans.
- The advisory committee shall revisit the Net Metering agreement and identify necessary modifications to the current small-scale solar PV regulations issued by ECRA.
- It is suggested to involves representatives of CSOs, solar energy SMBs and non-business stakeholders such as environmental and religious groups in periodic meetings with the advisory committee to assure their views are integrated into the polices design process.
- The advisory committee shall set identified short, mid, and long-term integration goals of RSPSs as a part of the country's energy portfolio.
- It is recommended to develop a national interactive platform, where interested individuals can assess the required upfront investment, estimated electricity output, required space, and potential revenue based on their dwelling location prior to the purchase of RSPSs.
- It is recommended to introduce an incentive scheme of dynamic pricing for uptake excess electricity (ToU) and take the necessary measures to maintain the service provider vested interests. Such a mechanism would encourage individuals to reduce their electricity consumption during peak times by trying to save excess electricity to export it to the grid and take advantage of the high uptake rates at these particular times. Such a mechanism is believed to help reduce the pressure of energy demand during peak times.
- The government shall grant a share of the subsidies allocated from conventional energy to introduce subsidizing programs and tax deduction to relieve the high upfront cost of RSPSs and encourage plausible integration of RSPSs.
- There is a need for further hikes to be applied to the electricity tariff in order to enhance the economic viability of RSPSs, taking into consideration the necessity to readjust the financial support programs dedicated to low and middle-class citizens (e.g., Citizen's Account Program).
- It is suggested to apply the new plan on a small-scale area or city to examine their effectiveness and competence before generalizing it to other cities.
- Besides the conventional methods, promotion campaigns shall be carried out by employing religious approaches.
- It is recommended to link country branding concepts to the undergoing efforts of developing solar energy megaprojects.
- It is suggested to further incorporating materials in the education system to increase awareness about the benefits of renewable energies.

Annex

i. **Distribution of the innovation attributes scores** (Fig. 6.1)

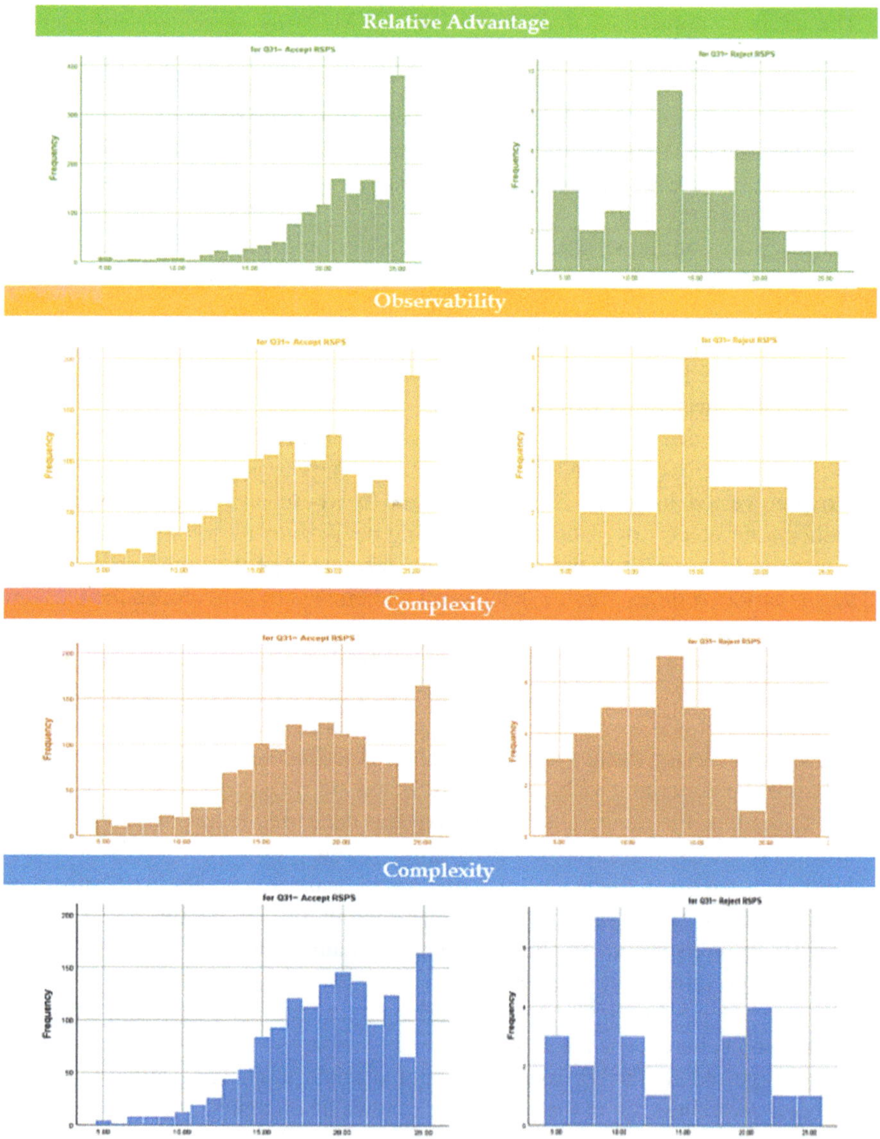

Fig. 6.1 Distribution of the innovation attributes scores among acceptors (left) and rejectors (right) of RSPSs

References

McKenna, E., & Thomson, M. (2014). Demand response behaviour of domestic consumers with photovoltaic systems in the UK: An exploratory analysis of an internet discussion forum. *Energy, Sustainability and Society, 4,* 13. https://doi.org/10.1186/2192-0567-4-13.

Moezzi, M., Ingle, A., Lutzenhiser, L., & Sigrin, B. O. (2017). A Non-Modeling Exploration of Residential Solar Photovoltaic (PV) Adoption and Non-Adoption (No. NREL/SR–6A20-67727, 1379469). https://doi.org/10.2172/1379469.

Correction to: An Exploratory Study of the Public's Views on Residential Solar Photovoltaic Systems in Hydrocarbon-Rich Saudi Arabia

Correction to:
Chapter 3 in: K. A. Alrashoud, *Managing Technology Transition in Saudi Arabia*, https://doi.org/10.1007/978-981-15-5803-0_3

The original version of the book was inadvertently published with an incorrect figure in Chapter 3, which has now been corrected as below. The erratum chapter and the book have been updated.

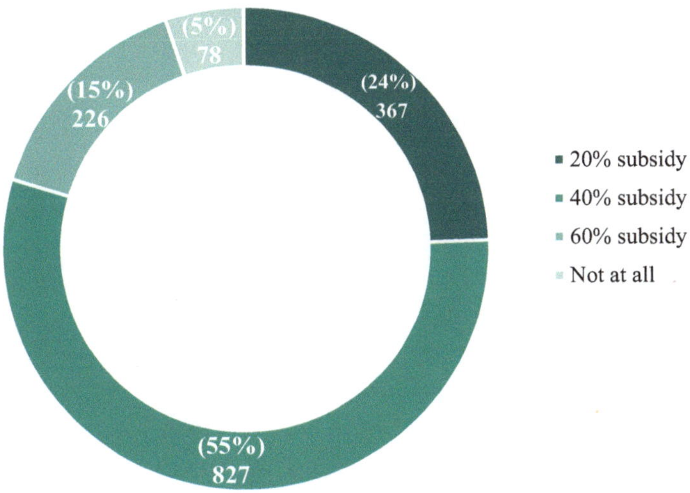

Fig. 3.7 Participants' willingness to install a residential solar photovoltaic system with subsidies

---

The updated version of this chapter can be found at
https://doi.org/10.1007/978-981-15-5803-0_3

K. A. Alrashoud, *Managing Technology Transition in Saudi Arabia*,
https://doi.org/10.1007/978-981-15-5803-0_7

Kolibri 360 GmbH
Ferdinand-Jühlke-Straße 7
99095 Erfurt, Deutschland
produktsicherheit@kolibri360.de